ATZ/MTZ-Fachbuch

MAHLE GmbH
Editor

# Cylinder components

## Properties, applications, materials

2nd Edition

Springer Vieweg

Editor
© MAHLE GmbH
Stuttgart, Germany

This book is based on the second, revised edition of the German book „Zylinderkomponenten" edited by MAHLE GmbH.

ATZ/MTZ-Fachbuch
ISBN 978-3-658-10033-9 ISBN 978-3-658-10034-6 (eBook)
DOI 10.1007/978-3-658-10034-6

Library of Congress Control Number: 2016933261

Springer Vieweg


Printed on acid-free paper

This Springer Vieweg imprint is published by Springer Nature
The registered company is Springer Fachmedien Wiesbaden
(www.springer.com)

# Preface

Dear readers,

This is the second, revised edition of the first volume of the MAHLE Knowledge Base, a multivolume set of technical books. This first volume, like the second volume "Pistons and engine testing," will make your daily work in this field of conflicting priorities somewhat easier and will be a good source of guidance for all the difficult questions, providing a good visual overview of the entire subject with many illustrations, charts, and tables. The MAHLE Knowledge Base is aimed at engineers and scientists in the areas of development, design, and maintenance of engines, at professors and students in the fields of mechanical engineering, engine technology, thermodynamics, and vehicle construction, and of course at any reader with an interest in modern gasoline and diesel engines.

The development and design of combustion engines is currently going through an extremely exciting phase. Never before have the demands of international lawmakers, customers, and consumer organizations been so contradictory, in part, in their effects on the design and development of engines. Environmental protection through clean exhaust gas, for instance, is not free of charge, neither in terms of costs, nor in terms of engine weight. Particulate filters, exhaust gas recirculation, SCR systems, and other solutions for exhaust gas treatment are also often in direct conflict with the goal of lower fuel consumption.

In this first volume, we present all the details of important cylinder components in meticulous scientific depth. Many questions concerning piston rings, piston pins and pin circlips, bearings, connecting rods, crankcases, and cylinder liners are answered. The contents reflect the experience, knowledge, and technical expertise of the engineers and scientists at MAHLE.

Many descriptive photos and graphics provide information on recent and future trends in cylinder components. Whether it is materials, types, coatings and surface treatments, numerical simulations and FE analyses, or casting processes; no relevant subject was left out. We wish you much enjoyment and many new insights from this reading.

Stuttgart, October 2015

*Wolf-Henning Scheider*
Chairman of the Management Board and CEO

*Heinz K. Junker*
Chairman of the Supervisory Board

# Acknowledgment

We wish to thank all the authors who contributed to this volume.

Dipl.-Ing. Juliano Avelar Araujo, Brazil
Dipl.-Ing. Beat M. Christen, Germany
Dipl.-Ing. Jürgen Dallef, Germany
Dipl.-Ing. André Ferrarese, Brazil
Dr.-Ing. Rolf-Gerhard Fiedler, Germany
B.Eng. James George, Great Britain
Dr. rer. nat. Roger Gorges, Great Britain
David Hancock, Great Britain
Dipl.-Ing. Daniel Hrdina, Germany
Michael Bernhard Hummel, Germany
CEng. MIMechE Mike Jeremy, Great Britain
Dipl.-Ing. Horst Kaiser, Germany
Dipl.-Ing. Oliver Kroner, Germany
Dipl.-Ing. Ditrich Lenzen, Germany
Dipl.-Ing. Roland Lochmann, Germany
Ing. Josef Locsi, Germany
Dr.-Ing. Daniel Lopez, Germany
B.Eng. Sebastian Mangold, Germany
Dipl.-Ing. Leandro Mileo Martins, Brazil
Günther Mayer, Germany
Dipl.-Ing. Marcelo Miyamoto, Brazil
Dipl.-Ing. Marco Maurizi, Germany
Dr.-Ing. Uwe Mohr, Germany
Dipl.-Ing. Eduardo Nocera, Brazil
Dipl.-Ing. Marcio Padial, Germany
Dipl.-Ing. Berthold Repgen, Germany
Dipl.-Ing. Andreas Seeger-van Nie, Germany
Dipl.-Ing. Anabelle Silcher, Germany
Dr.-Ing. Stefan Spangenberg, Germany
Peter Thiele, Germany
Dipl.-Ing. Adolf Tirler, Germany
Dr. Eduardo Tomanik, Brazil
Dipl.-Ing. Achim Voges, Germany
Dipl.-Ing. Oliver Voßler, Germany
Prof. Dr.-Ing. Stefan Zima (†), Germany

# Table of contents

# 1 Piston rings

## 1.1 Purpose and function of piston rings

Piston rings fulfill the following important tasks for engine operation:

- Sealing off the combustion chamber, in order to maintain the pressure of the combustion gas. The combustion gas must not enter the crankcase (also known as blow-by) and lubricating oil must not enter the combustion chamber.
- Transfer of heat built up in the piston to the cylinder surface
- Controlling the oil balance, where a minimum oil quantity needed to form a hydrodynamic lubricating film must reach the cylinder surface, while oil consumption needs to be kept as low as possible

The piston ring pack usually consists of three piston rings: two compression rings (also known as the first and second piston rings) and an oil control ring (third piston ring).

The piston rings perform the following functions:

1st piston ring: compression of combustion air or gas mixture, and support of gas pressure in the operating cycle, dissipation of generated heat to the cylinder surface (see also Section 1.3), and, to a slight degree, scraping of the residual oil from the cylinder surface

2nd piston ring: support of the remaining gas pressure due to blow-by past the first piston ring, throttling piston land pressures and control of pressure ratios in the ring belt, scraping of oil from and dissipation of generated heat to the cylinder surface

3rd piston ring: homogeneous distribution of the oil for lubrication of the piston group/cylinder bore tribological system and scraping of excess oil

The following issues, however, must be considered in the design of piston rings:

- Scuffing: partial seizure process leading to severe wear, poor sealing, increased oil consumption, and increased blow-by value
- Ring flutter or radial collapse: incidence of radial or axial instabilities that lead to leakage and therefore to increased blow-by
- Ring sticking: at excessive piston temperatures, the oil in the ring grooves carbonizes, so that the piston rings get stuck in it.
- High oil consumption: determining factors are the ring conformability (see Section 1.5.1.4) of the piston rings, deformation and honing of the cylinder bore, and the gas pressure ratios in the piston land region.
- Friction: the piston rings have a large part in the friction of the piston group.

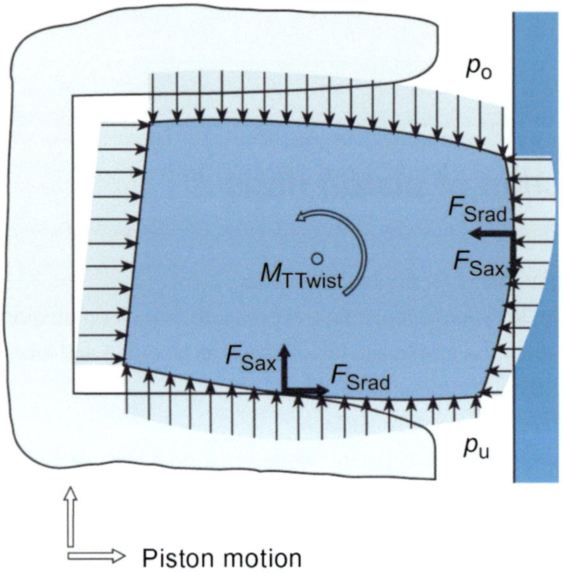


**Figure 1.1:**
Forces acting on a piston ring in the piston ring groove

Light blue: piston ring groove

Medium blue: piston ring

Dark blue: cylinder

Arrows encompassing the piston ring: forces acting on the piston ring

$p_o$: gas pressure above the piston ring

$p_u$: gas pressure below the piston ring

$F_{Srad}$: radial force and counterforce

$F_{Sax}$: axial force and counterforce caused by friction

$M_{T\ Twist}$: countermoment of the piston ring

The arrow shows the direction of positive twist.

Compression rings are mostly single-piece, with a spring force. Their basic shape is a thin-walled, axially short circular cylinder. To generate the necessary contact pressure against the cylinder wall, the piston rings are in the shape of an open circular spring. The spring force acting radially in the installed state is greatly amplified by the gas pressure behind the piston ring. Axial contact with the ring groove flank is substantially generated by the gas pressure applied to the piston ring side face (**Figure 1.1**).

When the piston is installed in the cylinder, the piston rings are compressed at their ends to their gap clearance. In the piston, they are guided in piston ring grooves corresponding to their dimensions and therefore follow the piston motion. This type, invented in 1854 by John Ramsbottom, is known as a self-tightening ring and has proved itself from the beginning in pistons for steam locomotives. It became a basic invention in engine technology, because reliable sealing of high gas pressures in the combustion chamber was first made possible by this type of ring–up to more than 260 bar today.

The force with which a piston ring presses against the cylinder wall depends mainly on the difference in diameters of the prestressed piston ring and the cylinder. This prestressing is designed in such a way that the piston ring meets the particular requirements arising from the working process and operating conditions. When the piston ring is installed in the cylinder, a tangential force is created that in turn generates the contact pressure.

- The radial distribution of the contact pressure is achieved by the shape of the piston ring–for example, by CNC turning or coiling.
- The radial distribution of the contact pressure depends on the shape of the running surface–cylindrical or conical–and the profile geometry of the piston ring (barrel shape).
- The contact pressure is determined by the working process.

The radial pressure applied by the piston ring to the cylinder bore is small in comparison to the gas pressure applied by the ring groove in the piston to the inner side of the piston ring (**Figure 1.1**). In diesel engines, with their high gas pressures, the piston ring is, in many cases, shaped against the running surface such that the gas pressure acts from here against the one on the inner side, which reduces the contact pressure on the cylinder surface. Owing to the ring gap dictated by the assembly process, the piston ring cannot provide complete sealing, which leads to leakage at this point.

Piston ring materials require

- good running and boundary lubrication capability;
- elastic behavior;
- mechanical strength;
- high strength at elevated temperatures;
- high heat conductivity; and
- good machinability.

Materials used include untempered and heat-treated gray cast iron, cast iron with nodular graphite (heat-treated), and tempered steel or nitrided stainless steel.

To improve running-in characteristics, reduce wear, and prevent scuffing, special measures are taken in coating and reinforcing (protecting) the running surfaces.

Operating behavior depends on many influence variables, which often makes the optimization of piston rings complex:

- Type and design of the engine
- Combustion process, combustion sequence, pressures, pressure gradients, aftertreatment technology, etc.
- Engine block and cylinder design, cylinder material and finishing (e.g., honing)
- Fuel and lubricant
- Piston technology
- Piston ring type, material, and running surface
- Operating conditions

## 1.2 Functional principles

As part of the moving boundary of the engine operating space, the piston ring fulfills various tasks. For the course of the thermodynamic process, it must ensure that the gas pressure in the cylinder is maintained and does not drop off. This is the task, in particular, of the first piston ring. One premise is that lubrication, acting as a "gas-sealing oil pressure barrier," is present. Tests by Felix Wankel had demonstrated that without such a fluid layer, higher gas pressures cannot be sealed against moving parts. The motion of the piston ring develops a hydrodynamic pressure that is greater than the gas pressure. This is why it is so important for the function of the piston ring that the cylinder surface is sufficiently wetted with lubricating oil. The main distribution of this oil quantity is performed by the oil control ring, while fine control is achieved by the first piston ring through oil control.

The arrangement of several piston rings in series forms a system of throttle chambers, in which the pressure of leaking gases is further decreased by throttling and swirling. It is unavoidable, however, that a small portion of combustion gases, compressed mixture, or air will pass by the piston rings and enter the crankcase (blow-by gas). The width and tolerance of the ring gap has a significant effect on the blow-by value. The piston ring seals against the side faces like a valve. Leakage points are most noticeable at the running surface, because the blow-by gas breaks through the oil film. In general, the blow-by value should be as low as possible, because the combustion gases cause increased oil aging and component wear. A certain blow-by value is desirable, however, in order to prevent oil transport into the combustion chamber.

## 1.3 Forces and stresses

### Forces and temperatures on piston rings

Piston rings are highly stressed mechanically, thermally, tribologically, and corrosively.

Piston rings must fulfill their task at high combustion gas temperatures and combustion pressures of up to 260 bar.

Depending on the design, up to 20% of the heat absorbed by the piston can be transferred to the cylinder wall by the piston rings.

The limit of the temperature load on the first piston ring is reached when the oil in the top ring groove starts to carbonize as a result of excessive temperature. The motion of the first piston ring, which is a requirement for its reliable function, is thereby limited. It can no longer maintain its proper contact with the cylinder surface, resulting in ring sticking. One ring-based solution is the keystone ring (**Figure 1.2**), developed in the early 1930s by the English engine manufacturer Napier.

Effective piston cooling is critical, as it significantly reduces the thermal load on the piston rings. Depending on the type of piston cooling, the heat flowing into the piston rings can be reduced.

During one revolution of the crankshaft, the piston moves from the top to the bottom (BDC) and back to the top dead center (TDC). It travels twice the stroke distance. During this

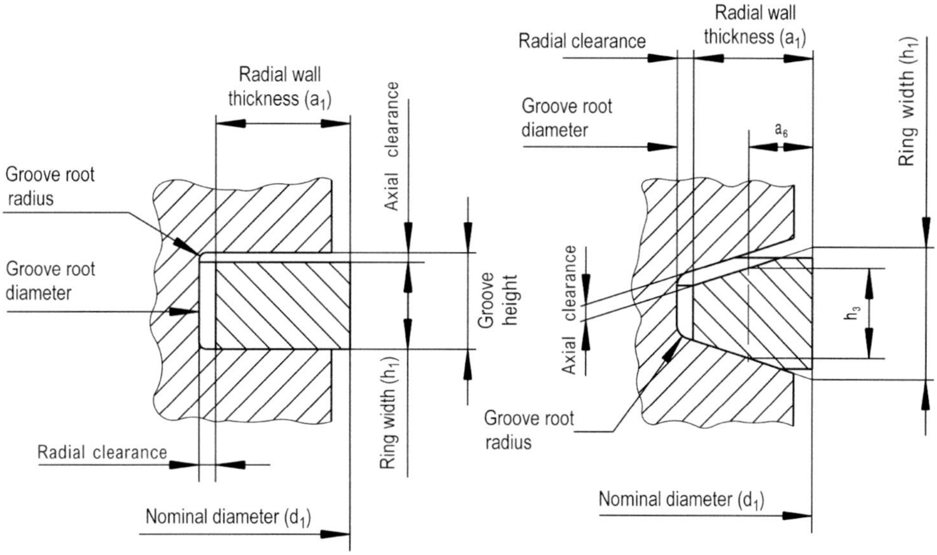


**Figure 1.2:** Rectangular ring (left) and keystone ring (right), axial clearances

motion, it is accelerated and decelerated. Owing to its inertia, the piston ring moves in the ring groove relative to the piston. Because of friction forces at the cylinder surface, it tends to tilt as it moves ( **Figure 1.1**). Upon impact, it can exert forces on the side faces of the ring groove. In diesel engines, this effect is increased further by the high gas pressure.

Wear on the groove flanks degrades the function of the piston ring, until it causes ring scoring, ring failure, and, as a result, piston seizure. The introduction of aluminum pistons for diesel engines used in commercial vehicles at the beginning of the 1930s nearly failed because of this type of damage, until Ernst Mahle created an effective solution with the ring carrier as a groove protector (**Figure 1.3**).

The high gas temperatures to which the first piston ring, in particular, is subjected, even if only for a short time, make its function more difficult, since together with the action of gas pressure, they impair the lubrication film between the first piston ring and the cylinder surface. This puts the first piston ring into a tribologically critical operating condition.

The piston rings, piston, cylinder surface, and lubricant form a tribological system, where all sliding partners are responsible for proper operation. For the piston ring, influence factors are the type, design features, tangential force, prestressing, material, and coating; for the piston, the type and materials, cooling, and constructive design details; and for the cylinder surface, it is the material, finishing (honing), and contour accuracy (see Chapter 5). The lubrication depends on the lubricant itself (base oil, additives, viscosity class), sufficient wetting of the running surface, and temperatures within the system.

**Figure 1.3:**
Ring carrier piston

Depending on the type of combustion and fuel quality, combustion gases contain corrosive components, the worst of which is sulfur dioxide ($SO_2$). Sulfur dioxide promotes corrosive wear of the cylinder surface, mainly in the region of the TDC. The ring running surface is also corroded. Poorer fuels (heavy fuel oils) used to run large-bore engines (medium-speed four-stroke and slow-running two-stroke engines) intensify this problem and require special measures on the ring, piston, and cylinder.

The motion of the ring pack generates friction and thus mechanical losses. Between 10 and 20% of the total engine friction power loss is caused by the ring pack. Friction is determined mainly by the following factors:

- Contact pressure (tangential force and gas pressure)
- Ring width
- Coefficients of friction of the contact surface (coating)
- Running surface design (barrel shape)
- Surface condition of the counterpart (cylinder surface)

Reduction of friction losses can be achieved primarily by minimizing contact pressure, i.e., by reducing the tangential force and ring width.

## 1.4 Types of piston rings

The various tasks of the piston rings can no longer by met by a single ring type. Thus, it made sense to classify the piston ring types in use today. This classification was made in DIN ISO 6621, Part 1, corresponding to **Figure 1.4**.

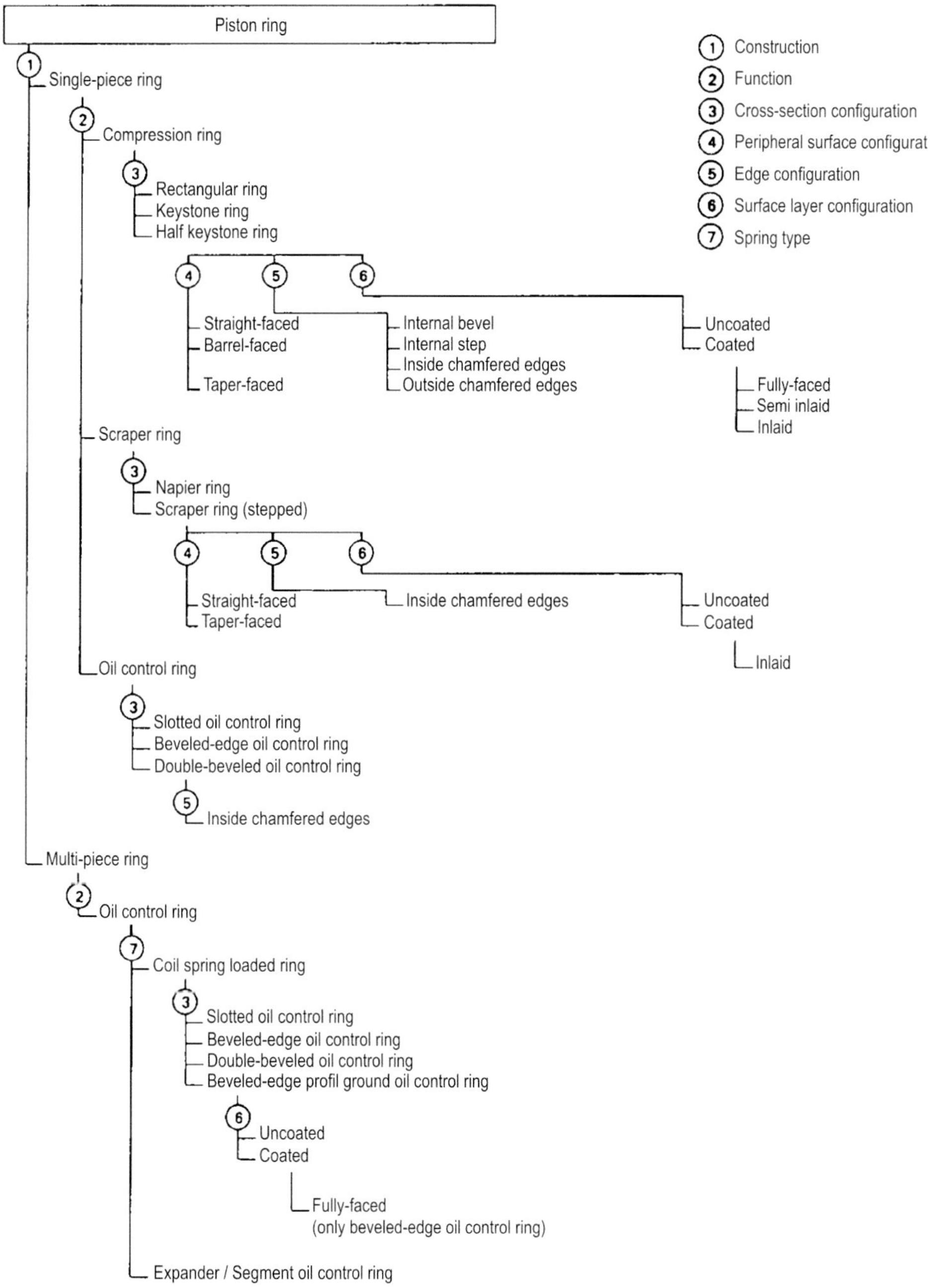


**Figure 1.4:** Classification of piston rings per DIN ISO 6621 Part 1, Section 4, p. 13

In recent years, the width of the piston rings has been drastically reduced. Today's compression rings in passenger car gasoline engines are typically 1.2 to 1.0 mm. For comparison: in the 1930s, the ring width was two to three times greater. Lower piston rings have lower mass, require less installation space, and allow a lower compression height of the piston. They also demonstrate better operating behavior in terms of friction, ring flutter, and blow-by. Precise machining of the piston ring grooves is therefore tremendously important. For extreme ratios of radial piston ring width to axial piston ring width, the piston rings become unstable.

Individual engine types–passenger car gasoline engines, passenger car and commercial vehicle diesel engines, as well as medium-speed four-stroke engines and slow-running two-stroke diesel engines–are fitted with piston ring packs where the overall efficiency is matched to the specific operating conditions by combining and matching different piston ring types.

The first piston ring is closest to the combustion chamber. This means that it is exposed to very high mechanical and thermal loads. In order to ensure good temperature resistance, nodular cast iron or steel materials are used as the base material in these piston rings. They are also coated or specially treated, in order to reduce friction and wear. Piston rings are allowed to cause only minimal wear on the cylinder bore.

The first piston ring for commercial vehicle diesel engines subject to high stress generally has a keystone shape (see Section 1.4.5.). The running surface can be barrel-shaped and either symmetrical or asymmetrical (see Section 1.4.6). Asymmetrical profiles can reduce radial wear and improve oil consumption.

Even if the squareness of the ring groove has slight deviations, the piston ring remains in its line of contact with the cylinder surface. When the piston ring changes direction at the end of the stroke, contact is maintained between the running surface of the piston ring and the cylinder. Barrel-shaped piston rings cause less wear in the region of the cylinder surface, where the first piston ring changes its running direction. The barrel-shaped piston ring can be furnished with an internal bevel on its top edge, in order to achieve a positive distortion (also known as positive twist, see **Figure 1.1**). Strict requirements regarding lubricating oil consumption, however, have led to the first piston ring taking on part of the oil control task as well. In this regard, the running surface is given an asymmetrical barrel shape. Owing to the asymmetry, the center of the barrel shape is shifted in the direction of the lower half of the ring width. This improves engine run-in and oil control.

The second piston ring has a double function, depending on its type: it must seal against gas pressure while stripping oil off the cylinder wall; at the same time, sufficient lubrication of the first piston ring must be ensured. The second piston ring features a reinforced design with regard to its stripping effect, based on its additional function as an oil control ring. Its effectiveness is based on the contact pressure, the shape of the stripping surface (land), and the method of removal of stripped oil. This requires good ring conformability, i.e., the ability to adapt as smoothly as possible to the continuously changing cylinder deformation while maintaining the required contact pressure against the cylinder wall. Friction and wear need to be kept to a minimum.

### 1.4.1 Rectangular ring

The basic shape of the first piston ring is a rectangular ring with a cylindrical running surface, also known as an R-ring (**Figure 1.5**). Its task is to seal against the gas pressure in the combustion chamber. Rectangular rings are used for normal operating conditions, primarily as first piston rings in gasoline engines.

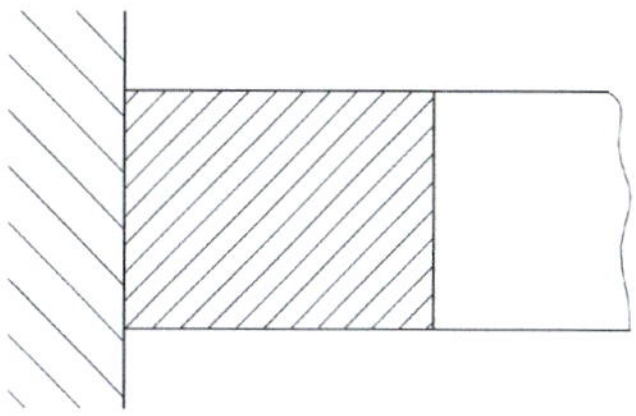

**Figure 1.5:** R-ring

### 1.4.2 Rectangular ring with conical running surface

A slight taper (conicity) to the outer surface of the piston ring increases its effectiveness. Contact between the piston ring and the cylinder wall is reduced to a narrow line. This line contact increases the contact pressure of the piston ring against the cylinder bore and ensures that contact is maintained with the bore, even if the cylinder is deformed. The run-in phase is thereby shortened. It also provides a downward stripping effect, which supports the oil control function of the oil control rings. This type of ring, also called a taper-face ring or M-ring, is typically employed as a second piston ring (**Figure 1.6**).

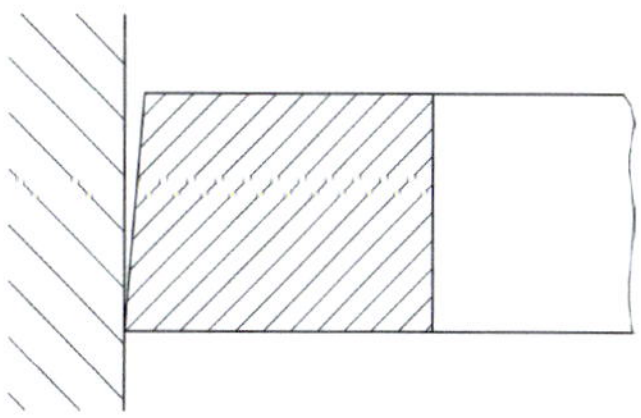

**Figuro 1.6:** M ring

### 1.4.3 Piston ring with internal bevel or internal step (top)

Because of a chamfer on the top inner side of the piston ring (internal bevel IF), the forces of the piston ring are modified such that its cross section tilts about its axis, as a result of compression during installation of the piston in the cylinder.

This distortion (twist, i.e., a tilted position of the piston ring under tension) provides a line contact of the bottom oil scraper edge against the cylinder surface, as well as between the piston ring side face and the piston groove flank. The latter reduces the passage of combustion

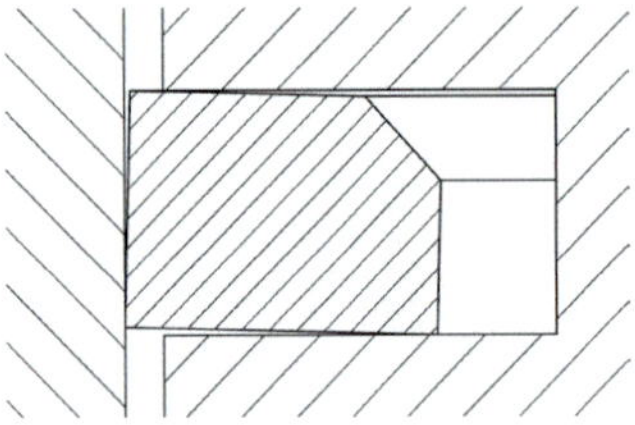

**Figure 1.7:** R-ring with top internal bevel

gases as well as engine oil. When the internal bevel is at the top, it is referred to as a positive twist. Taper-face rings (second ring) can also be designed with a positive twist. In the past, this design was used as a measure for further reducing blow-by.

Piston rings of this type, also known as R-rings with top IF, are used both as first and as second piston rings (**Figure 1.7**).

### 1.4.4 Piston ring with internal bevel or internal step (bottom)

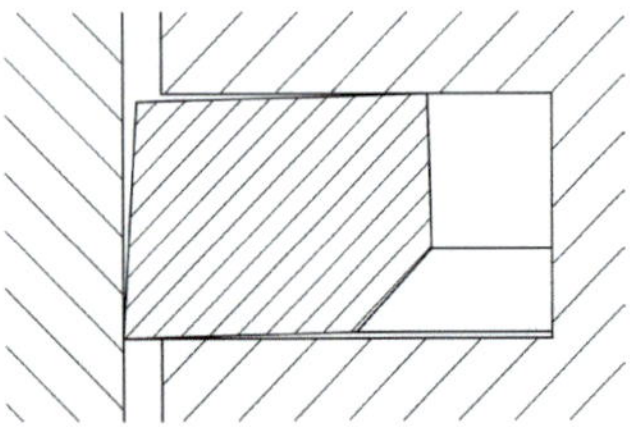

**Figure 1.8:** M-ring with bottom internal bevel

In contrast to Section 1.4.3, moving the internal bevel to the bottom provides a negative twist. These piston rings with bottom internal bevel (IFU), also called M-rings with bottom IFU, make contact at the bottom with the cylinder and at the top inside with the groove flank (**Figure 1.8**). Such piston rings are preferably installed in the middle ring groove and are part of the group of oil control rings. With regard to oil control, contact of the lower part of the running surface against the cylinder surface is desired. Oil control rings with greater conicity are therefore used to compensate for the twist. The negatively twisted piston ring creates a good seal at the bottom against the cylinder surface, thanks to its linear contact, and prevents oil from entering the ring groove. This is especially important for low pressures in the combustion chamber, such as can occur when the mixture is throttled in gasoline engines or at gas exchange. In addition, a second ring with a negative twist can bring about a controlled axial motion in order to control the pressure ratios in the second groove and thus the oil transport mechanisms. The superior oil control of the negative twist in the second piston ring comes at the cost of slightly higher blow-by rates. The high gas pressures under full load deform both types of twisted piston rings in such a way that they are nearly flat at the bottom groove flank. Under partial load, the piston ring deformation is not as severe, making the behavior of the rings more effective.

### 1.4.5 Keystone ring

Keystone rings are divided into half and full keystone types. On a half keystone ring, also known as an HK-ring, only one side has a conical design; on a full keystone ring, also known as a K-ring, both sides do (**Figure 1.9**).

Figure 1.9: HK-ring (left) and K-ring (right)

These piston ring geometries reduce carbon buildup in the ring groove. The radial motion of the piston ring in the ring groove keeps it clear of oil carbon. Keystone rings of both types are mostly used as first piston rings in commercial vehicle diesel applications.

### 1.4.6 First piston ring with barrel-shaped surface

In the early days of engine technology, it was commonly believed that the first piston ring would seal even better, the more precisely it matched the geometric rectangular shape. Despite great effort to obtain the greatest dimensional accuracy in manufacturing, the operating performance of the first piston ring did not improve; rather, it got worse. Practical experience demonstrated that the sealing behavior of the first piston ring improved over time, when the sharp square corners had been worn off. This wear state was then anticipated, first by chamfering, then with a barrel-shaped running surface.

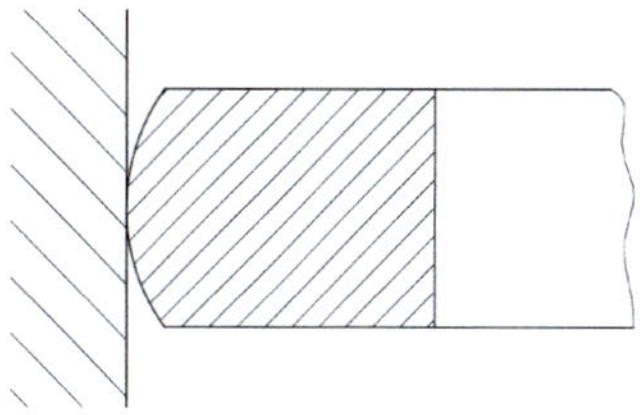

Figure 1.10: R-ring B

With the barrel shape, better hydrodynamic lubricating conditions are achieved, and the axially shorter contact surface at the cylinder surface improves sealing. In addition, the negative effects of cylinder deformations during engine operation can be better compensated. Piston rings of this type, also known as R-ring B, are used as first piston rings (**Figure 1.10**).

### 1.4.7 Napier ring with conical running surface

Thanks to a conical running surface, the run-in period of the taper-faced Napier ring is shortened and its oil-stripping effect is amplified. The hook of the taper-faced Napier ring acts as an oil reservoir for scraped oil and prevents it from entering the ring groove. This type of design, also known as the NM-ring, is used as a second piston ring (**Figure 1.11**).

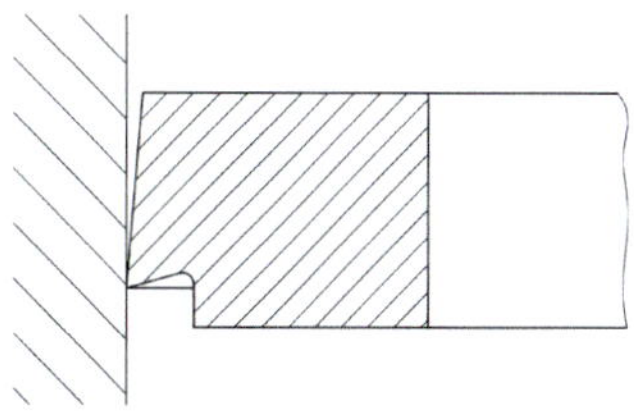

Figure 1.11: NM-ring

## 1.4.8 Ring gap configuration

The gap of the piston rings generally has a straight shape. Other types of gaps are used in engines for special requirements.

In two-stroke and opposed-piston engines, in which rotation of the piston rings is undesired, an inner or flank recess is made in the ends of the ring, where a safety dowel pin is located in the piston. This secures the piston ring in its location in the piston, which prevents damage to intake and exhaust slits and to the ring ends in two-stroke engines (**Figure 1.12**). In an opposed-piston engine, this prevents the ring gaps from all being located at the same place on the piston circumference, which would produce increased blow-by, for example.

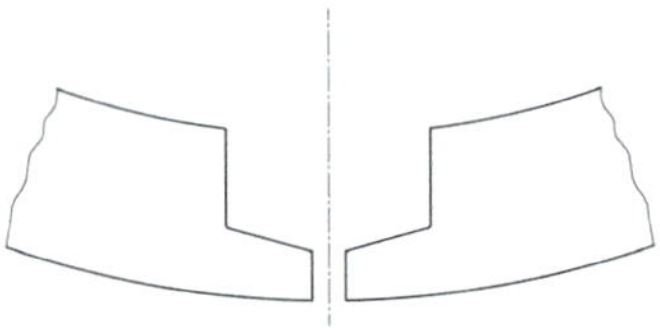

**Figure 1.12:** Flank recess (left) and inner recess (right)

Rings that are meant to seal rotating shafts and for which the piston ring side face acts as a sealing element are designed with an overlapping joint (only for uncoated piston rings) (**Figure 1.13**). Another alternative is the piston ring with an interlocking joint (only for uncoated piston rings).

For high blow-by quantities, a taper-faced Napier ring is employed in the middle ring groove (see Section 1.4.7) with a gap in the groove. The gap in the groove at the joint reduces the passage of combustion gases.

**Figure 1.13:** Overlapping joint (left) and interlocking joint (right)

## 1.4.9 Slotted oil control ring

The slotted oil control ring is a single-piece ring, which contacts the cylinder surface with two lands. Penetrations are cut into the middle area of the ring body between the two lands, acting as oil drainage points and ensuring good conformability of the ring. The smaller total contact surface (land width) increases the contact pressure against the cylinder surface. This is necessary, as no gas pressure can build up behind the slotted oil control ring. The contact pressure of the oil control rings thus arises from their tangential force. Further reduction in the size of the land surfaces resulted in the beveled ring (D-ring), with chamfers on the lands, and the double-beveled oil control ring, with uniformly aligned chamfers on the lands (G-ring) (see Section 1.4.10.1, but here with coil springs). Single-piece oil control rings are assembled in the bottom ring groove, but are seldom used in original equipment manufacturer applications, where the majority of the designs use spring loading.

## 1.4.10 Spring-loaded oil control ring

### 1.4.10.1 Oil control ring with coil spring

To improve ring conformability and homogeneously distribute the contact pressure, two-piece oil control rings are preloaded with a cylindrical spring (coil spring) on the inside of the ring (SSF-ring). The ends of the spring support each other (**Figure 1.14**).

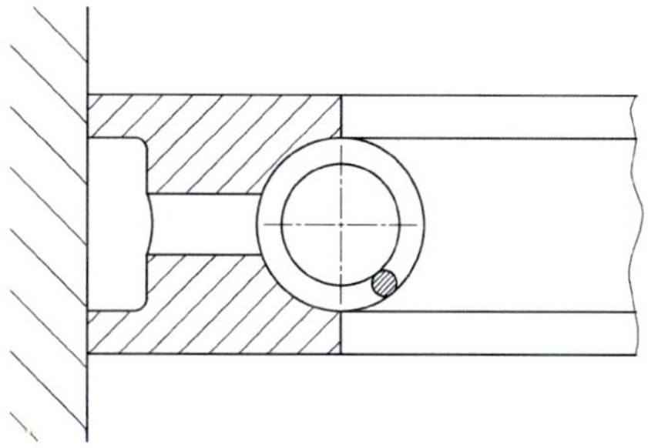

**Figure 1.14:** SSF-ring

Owing to the flat characteristic curve, the spring preload changes very little, even after long periods of operation. Narrow (axially low) piston rings are intended to improve ring conformability. Smaller piston ring axial widths also have a direct effect on the compression height, and thus on piston weight, with all the associated advantages. Typical oil control ring axial widths for diesel applications range between 2.0 and 3.5 mm, depending on the application (passenger car, commercial vehicle). As with springless piston rings, there is a beveled ring with coil spring (DSF-ring) and a double-beveled oil control ring with spring (GSF-ring) (**Figure 1.15**).

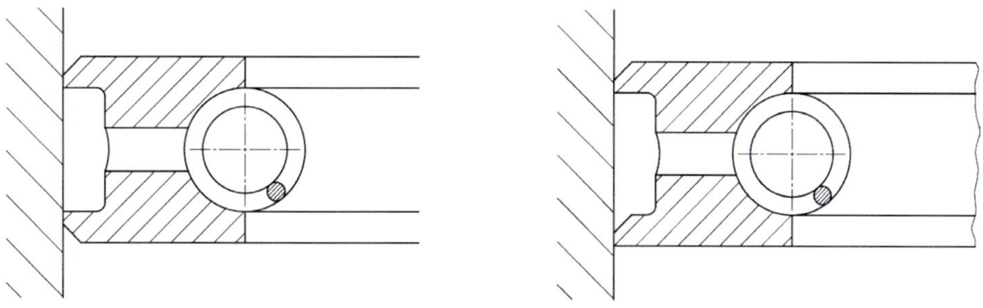

**Figure 1.15:** DSF-ring (left), GSF-ring (right)

In most gasoline engines, three-piece oil control rings are used mainly for cost reasons and on account of their axial sealing capability in the partial-load range. In view of their required engine service life, diesel engines also require higher durability, which can normally be achieved more easily with two-piece oil control rings.

One of the most important characteristics of oil control rings is the specific contact pressure. Overall, the consumption of lubricating oil is lower, the higher the specific contact pressure (because of better oil control). In order to reduce consumption of lubricating oil during engine run-in, a taper can be applied to both contact lands. The angled running surface reduces the contact zone with the cylinder surface, thus providing greater contact pressure during run-in, which reduces the normally higher lubricating oil consumption in this stage. After a certain running time, the angled profiles wear down and take on a cylindrical shape.

### I-shaped oil control ring

The I-shaped oil control ring is a two-piece design and uses steel as the base material (Figure 1.16). In contrast to oil control rings made of cast iron, these rings are produced from a preformed steel wire with an I-shaped cross section. This is coiled, cut to length in the appropriate shape, and then finish machined. In order to increase wear resistance, the I-shaped oil control rings are usually nitrided.

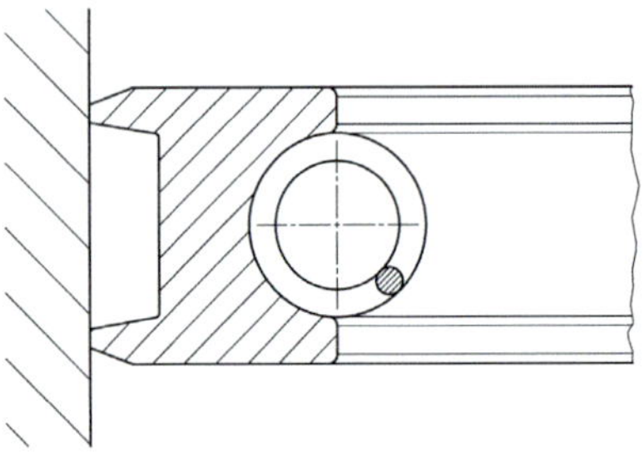

**Figure 1.16:** I-shaped oil control ring made of steel

I-shaped oil control rings are recommended particularly for high-speed diesel engines, as well as for highly stressed diesel engines, which are expected to last at least one million kilometers in commercial vehicles. In special cases, they are also used in high-performance gasoline engines. This piston ring design is also used as an oil control ring in the bottom ring groove.

One step toward reducing tangential force and thus friction power loss with two-piece oil con-

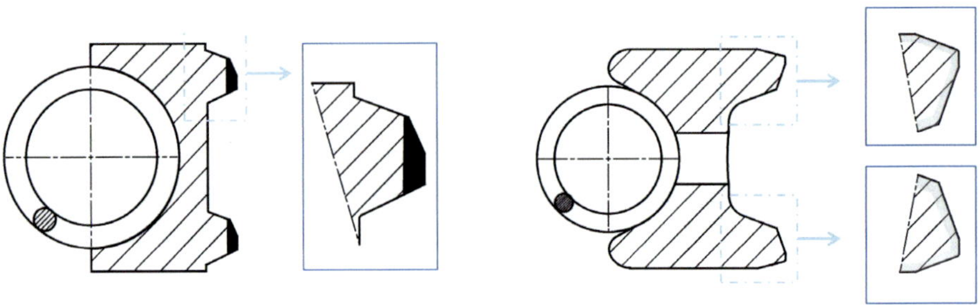

**Figure 1.17:** X-taper design (left) and V-shape design (right) for oil control rings with optimized friction power losses

trol rings, while maintaining sufficient oil control, has been the development in recent years of new land designs that have led to additional reductions in land width. Examples include the MAHLE X-taper or V-shape designs, which combine a small land width (less than 0.15 mm) with a large taper angle to reduce the influence of wear (**Figure 1.17**).

### 1.4.10.2 Three-piece oil control ring (expander ring)

**Three-piece steel ring (3-S-ring)**

It is made of two steel rails that are held in position by a spacer spring and are radially preloaded. The running surfaces of the rails are typically coated (e.g., chrome-plated, nitrided, or PVD-coated) to protect against wear. The spring is part of the load-bearing piston ring construction (**Figure 1.18**).

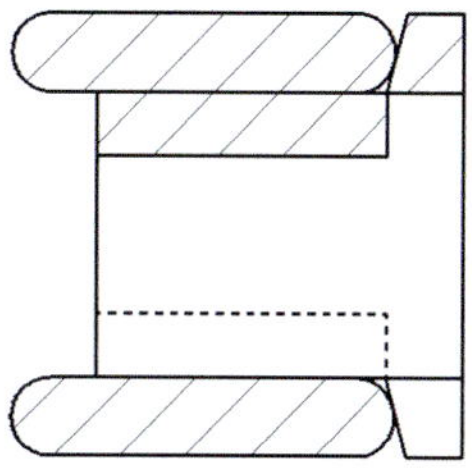

**Figure 1.18:** 3-S-ring

The rails strip off the excess oil from the cylinder surface. There are different types of three-piece oil control rings. Their functional principle is substantially the same, namely, two steel rails are pressed against the cylinder wall by an expander. These expanders of varying shape must fulfill the following tasks: they need to press the rails against both the cylinder surface and the groove flanks, and thus seal them off. Oil entering between the two rails is returned to the crankcase.

Oil penetration into the combustion chamber from the piston ring groove is reduced. The oil collected between the rails can also enter the piston interior through slots.

Such piston rings are often used as third piston rings in gasoline engines. This is mainly for cost reasons, but also because of the oil consumption benefits in the partial-load range due to good lateral (axial) sealing of the piston ring groove with the rails.

## 1.4.11 U-flex ring

The U-flex ring is a one-piece, closed ring whose ends touch. The ring is made of elastic spring steel. It is stamped, then bent into a U-shape and coiled (**Figure 1.19**). The U-flex ring is generally installed with a coil spring (for assembly purposes only).

Its special shape and manufacture give the U-flex ring very good properties with regard to ring conformability, allowing good oil control with low tangential forces, and therefore low friction. Its good ring conformability makes the U-flex ring very well suited for engines with higher-order bore deformations.

Today, the U-flex ring is used in both gasoline and high-speed diesel engines.

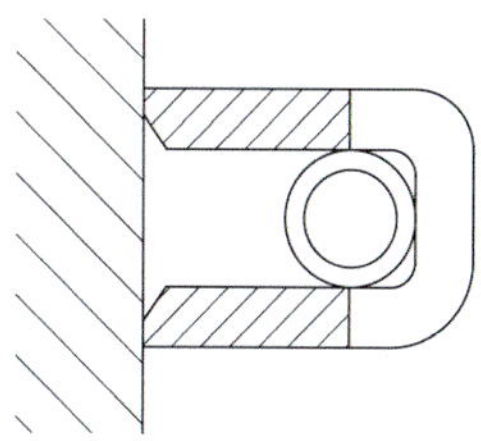

**Figure 1.19:** U-flex ring

# 1.5 Design details

## 1.5.1 Analysis and simulation

### 1.5.1.1 Numerical analysis

The design of new piston rings and creation of design and production drawings is based on databases in which all the important dimensions and properties are collected and stored. On the basis of these files, which are continuously updated, piston rings are drawn directly using computer-aided engineering (CAE). In addition to dimensions, piston ring drawings also contain certain functional characteristics, such as the specific contact pressure, tangential force, and cross section of the piston ring.

### 1.5.1.2 Stress analysis

Piston rings are subjected to the greatest stress during installation, when they are stretched over the piston. The installation stress during the expansion ($S_a$) needed for assembly and the stress that arises in the cylinder in the installed state ($S_w$) can be calculated as follows:

$$S_w = \frac{8}{3 \cdot \pi} \cdot \frac{E \cdot t_y \cdot (m - s_1)}{(d_1 - a_1)^2}$$
$$S_a = \frac{8}{3 \cdot \pi} \cdot \frac{E \cdot (a_1 - t_y) \cdot (m_1 - m)}{(d_1 - a_1)^2} \quad (1-1)$$

$S_w$: stress in installed state
$S_a$: installation stress (expansion for assembly)
E: Young's modulus of the piston ring material
$t_y$: radial distance from the neutral axis to the ring running surface
m: free gap in relaxed state
$s_1$: gap clearance in installed state
$d_1$: nominal diameter of cylinder liner
$a_1$: radial dimension of piston ring
$m_1$: installation opening (normally, $m_1 = 9 \cdot a_1$)

For complex piston ring cross sections, such as two-piece oil control rings, the stresses are typically determined by finite element analysis.

### 1.5.1.3 Dynamic analysis

Using a numerical simulation, it is possible to analyze the interplay of piston rings, piston, and cylinder. The piston ring pack can be optimized, for example, with regard to blow-by and reduction of lubricating oil consumption. Such analyses are composed of

- thermal FE analysis of the cylinder;
- thermal FE analysis of the piston;
- computation of piston and piston ring dynamics;
- simulation of the engine cycle.

#### 1.5.1.4 Ring conformability

In the course of an operating cycle, the heat flow changes, which results in high temperature and pressure gradients in the piston and the cylinder liner. Together with the peak cylinder pressure in the combustion chamber and the assembly-induced stresses in the engine block, this leads to various distortions in the cylinder bores. The piston ring needs to adapt to these deformations, in order to keep blow-by and oil consumption low.

The ability of a ring to compensate for deformation can be expressed indirectly and in a simplified form by the coefficient k.

$$k = \frac{F_t \cdot (d_1 - 2 \cdot t_y)^2}{4 \cdot E \cdot I} \qquad (1-2)$$

k: coefficient of conformability
$F_t$: tangential force of the piston ring
I: axial moment of inertia of the piston ring cross section

The greater the value of the coefficient k, the better the conformability of the piston ring. The ability of a piston ring to make contact with the cylinder surface can be estimated as follows, according to Tomanik:

$$U_{max} = \frac{k \cdot d_1}{10 \cdot (i^2 - 1)} \qquad (1-3)$$

$U_{max}$: maximum cylinder deformation that the piston ring can adapt to
i: order of deformation (i = 1,2,3...)

#### 1.5.1.5 Specific contact pressure

One of the most important parameters is the specific contact pressure or unitary pressure. This is especially true for oil control rings. The specific contact pressure $P_0$ of the piston ring is derived from:

$$P_0 = \frac{2 \cdot F_t}{d_1 \cdot h_1} \qquad (1-4)$$

$P_0$: specific contact pressure
$h_1$: width of the piston ring

The high peak cylinder pressure (PCP) bears on the first and, to a lesser extent, the second piston ring, but dissipates during the operating cycle. For oil control rings, the ring width is replaced by twice the land width (two-piece oil control ring) or by twice the rail width (three-piece oil control ring).

#### 1.5.1.6 Ovality

Ring ovality is the maximum change of the nominal diameter of the piston ring, measured in various directions. It is determined by subtracting the diameter in the 90° and 270° direction from the direction of the stressed state.

#### 1.5.1.7 Design specifications

Piston rings are standardized with regard to their dimensions and properties. Nevertheless, adaptation of the piston ring design to the particular installation and operating conditions is often required.

## 1.6 Materials, coatings, and surface treatment

### 1.6.1 Materials

MAHLE has a complete range of piston rings made of gray cast iron, alloyed cast iron, and nodular cast iron, which are produced using cutting-edge casting. Carbon and stainless steel wire are obtained from leading global suppliers. The critical criteria for material selection are cost-effectiveness and engine specifications.

#### 1.6.1.1 Cast iron

For many years, lamellar cast iron with low alloying element content–but rich in graphite–was the suitable piston ring material. Its wear resistance, good running properties, mechanical strength appropriate for this purpose, as well as advantageous compatibility with cylinder liner and piston materials made it the optimal material for piston rings.

For a long time, cast iron was produced in single and double casting processes, which gave the material an attractive "A-class" graphite structure. With advancements in engine development, more complex piston ring materials with improved mechanical strength and wear resistance became necessary. Systematic developments in this area led to new types of alloyed gray cast iron and nodular cast iron.

MAHLE produces these materials in its own foundries with modern furnaces, in which the melt parameters are strictly controlled, which enables the manufacture of a wide range of first-class cast iron types.

The standard material MF 013 (perlitic lamellar cast iron, MC 13 according to ISO) is used for oil control rings in gasoline and diesel engines. The piston ring running surface is typically coated with chromium or another suitable material. The perlitic basic microstructure of the material and the uniformly developed lamellar graphite structure are excellent characteristics for a piston ring material that keep wear to a low level in uncoated oil control rings for gasoline engines.

In cases, where greater wear resistance is required, it is recommended that an alloyed material such as MF 025 (MC 25 according to ISO) be used. The material MF 032 (MC 32 according to ISO) can be used for applications with even higher requirements.

Alloyed types of cast iron are heat-treated in order to develop their mechanical properties. The resulting microstructure is primarily martensitic.

The mechanical properties of the nodular cast iron MF 053 (MS 53 according to ISO) are between those of gray cast iron and steel, although its self-lubricating properties are not as good as those of gray cast iron. This material is recommended for coated or uncoated compression and oil control rings, where the required strength is greater than that of lamellar cast iron.

For applications in which greater wear resistance is needed, in combination with the higher mechanical strength of nodular cast iron, the material MF 056 (nodular cast iron alloyed with niobium, MC 56 according to ISO) is recommended.

#### 1.6.1.2 Steel

Steel can be used to manufacture many types of piston rings, from the first to the third piston ring. These can be coated or nitrided piston rings, expanders, and rails of three-piece oil control rings, or I-shaped piston rings and springs of two-piece oil control rings. Steel is used in place of gray cast iron for its high mechanical strength and fatigue resistance, heat resistance, and good corrosion resistance. Steel rings are normally coated and/or nitrided.

### 1.6.2 Coatings and surface treatments

MAHLE piston ring coatings and surface treatments provide improved wear resistance and seizure resistance, along with low cylinder wear and favorable lubrication properties. Nanotechnology processes are also employed in this connection. Nitrided steel and cast iron, coatings based on chromium such as hard chrome and chromium-ceramic, plasma-sprayed molybdenum, plasma-sprayed cermet, and coatings using high-speed flame spraying (High Velocity Oxygen Fuel, HVOF) and physical vapor deposition (PVD) meet the most demanding service life and run-in requirements. The selection of a suitable coating depends on the engine technology, the application, the tribological requirements, and not least the cost.

Surface protection coatings and treatments intended to provide good oxidation resistance, such as tin-plating, black oxiding, ferroxidation, and phosphating, are available for specific applications.

#### 1.6.2.1 Gray cast iron as a base material

**MF 012**

Perlitic gray cast iron

Alloying elements: Cr, Cu

ISO 6621-3: Subclass 12

Second piston ring and two-piece oil control rings

Bending strength: min. 380 MPa

Hardness: 95 to 108 HRB

**MF 013**

Perlitic gray cast iron

Alloying elements: Cr, Cu

ISO 6621-3: Subclass 13

Standard material for compression and oil control rings in gasoline and diesel engines

Bending strength: min. 420 MPa

Hardness: 97 to 108 HRB

**MF 025**

Martensitic alloyed gray cast iron

High wear resistance

Alloying elements: Mo, Nb, V, W

ISO 6621-3: Subclass 25

High fracture strength with good wear resistance for second compression rings in gasoline and diesel engines

Bending strength: min. 650 MPa

Hardness: 37 to 45 HRC

**MF 032**

Martensitic carbidic gray cast iron

High wear resistance

Alloying elements: Mo, Nb, V, W

ISO 6621-3: Subclass 32

High fracture strength with good wear resistance for second compression rings in gasoline and diesel engines

Bending strength: min. 650 MPa

Hardness: 35 to 45 HRC

#### 1.6.2.2 Martensitic nodular cast iron as a base material

**MF 053**

Martensitic nodular cast iron

Alloying elements: Ni, Mo

ISO 6621-3: Subclass 53

First piston ring with high fracture strength and two-piece oil control rings with low land width in gasoline and diesel engines

Bending strength: min. 1,300 MPa

Hardness: 28 to 42 HRC

**MF 056**

Martensitic carbidic nodular cast iron

Alloying elements: Ni, Mo, Nb

ISO 6621-3: Subclass 56

First piston ring with high fracture strength and wear resistance

Bending strength: min. 1,300 MPa

Hardness: 35 to 45 HRC

#### 1.6.2.3 Carbon and stainless steels

**MS 068**

Carbon steel

Martensitic heat-treated

ISO 6621-3: Subclass 68

Base material for chrome-plated rails in three-piece oil control rings in gasoline engines

Tensile strength: no fracture in bending test

Hardness: 68 to 72 HR30N

**MS 067**

Austenitic stainless steel

Alloying elements: Cr, Ni

ISO 6621-3: Subclass 67

Expander ES-1 (type 81) for three-piece oil control rings in gasoline engines

Tensile strength: no fracture in bending test

Hardness: 59 to 67 HR30N

**MS 062**

Steel alloyed with chromium and silicon

ISO 6621-3: Subclass 62

Heat-resistant springs in two-piece oil control rings in diesel and gasoline engines

Tensile strength: 1,800 to 2,000 MPa

**MS 066**

Martensitic stainless steel

Alloying elements: Cr, Mo

ISO 6621-3: Subclass 66

Base material for nitrided, chrome-plated, or molybdenum-coated first piston rings in diesel and gasoline engines

Tensile strength: 1,125 to 1,325 MPa

Hardness: 38 to 42 HRC

**MS 064**

Steel alloyed with chromium and silicon

ISO 6621-3: Subclass 64

Base material for chrome-plated, molybdenum-coated, and high-speed flame-sprayed first piston rings in diesel and gasoline engines

Tensile strength: 1,590 to 1,960 MPa

Hardness: 48 to 54 HRC

#### 1.6.2.4 Running surface and side face coatings

**MCR 024**

Hard chrome plating

Galvanically applied

Piston rings in gasoline or diesel engines

Good wear resistance and seizure resistance

Hardness: min. 800 HV 0.1

**MCR 236/MCR 256**

Chromium-ceramic coating with $Al_2O_3$ (MCR 236) or cBN particles (MCR 256)

Galvanically applied

Piston rings in diesel engines

Excellent wear resistance and seizure resistance

Hardness: 900 to 1,200 HV 0.1

**MSC 278/MSC 280**

Mo + NiCr cermet alloys

Plasma-sprayed coatings

Piston rings in gasoline or diesel engines

Good wear resistance and high seizure resistance

Hardness: min. 450 HV

**MSC 380/MSC 385**

HVOF cermet coatings applied by high-speed flame spraying
For first piston rings in diesel engines
Superior wear resistance and seizure resistance
Hardness: min. 500 HV

**MIP 230/MIP 240/MIP 290/MIP 300**

Chromium-nitride coating (MIP 230/MIP 240)
Chromium nitride/nanobium nitride multilayer system (MIP 290)
Chromium carbon nitride coating (MIP 300, "CERAMSLIDE")
Coatings applied using physical vapor deposition (PVD)
For first piston rings in gasoline and diesel engines, as well as I-shaped oil control rings
Superior wear resistance and seizure resistance
Hardness: 1,200 to 1,600 HV (MIP 230); 800 to 1,200 HV (MIP 240); 1,700 to 2,100 HV (MIP 290); 1,800 to 2,200 HV (MIP 300)

**MIP 274**

Carbon-based coating (Diamond Like Carbon, DLC)
Coating applied using plasma-enhanced chemical vapor deposition (PECVD)
For first piston rings in gasoline and diesel engines
Superior seizure resistance and very good running-in characteristics

#### 1.6.2.5 Nitriding running surfaces

**MS 065 – N**

Nitrided 10 or 13% chromium stainless steel
Rails in three-piece oil control rings
High wear resistance
ISO 6621-3: Subclass 65
Hardness: min. 900 HV 0.050 at 0.01 mm, min. 700 HV 0.1 at 0.03 mm

**MS 066 – N**

Nitrided 17% Cr martensitic stainless steel
First piston ring in diesel engines, oil control rings in diesel and gasoline engines
High wear resistance
ISO 6621-3: Subclass 66
Hardness: min. 900 HV 0.050 at 0.01 mm, min. 700 HV 0.1 at 0.03 mm

**MS 067 – N**

Nitrided austenitic stainless steel

Expander ES-2 (type 81) in gasoline engines

Excellent heat resistance and low tangential force loss

ISO 6621-3: Subclass 67

Nitrided area: min. 0.004 mm

### 1.6.2.6 Surface protection

Some surface treatments can be used for special purposes, such as for oxidation resistance or for protection against microwelding (**Table 1.1**).

**Table 1.1:** Properties and applications of various protective coatings

| MCA standard | Protective coating or treatment | Groove | Properties |
|---|---|---|---|
| MPR 022 | Black oxiding | Oil control rings and rails | Oxidation resistance |
| MPR 023 | Manganese phosphate | First piston rings and oil control rings | Oxidation resistance |
| MPR 027 | Zinc phosphate | First piston rings and oil control rings | Oxidation resistance |

# 2 Piston pins and piston pin circlips

## 2.1 Function of the piston pin

The piston pin is the link between the piston and the connecting rod. Owing to the oscillating motion of the piston and the overlay of gas and inertial forces, it is subjected to high loads in alternating directions. **Figure 2.1** shows the piston pin load for a gasoline engine at rated power. The rotational motion of the connecting rod relative to the piston must be compensated for at the bearing locations of the piston pin, in the piston pin boss, and the small end bore. Because of the small relative motions, the lubrication conditions here are poor.

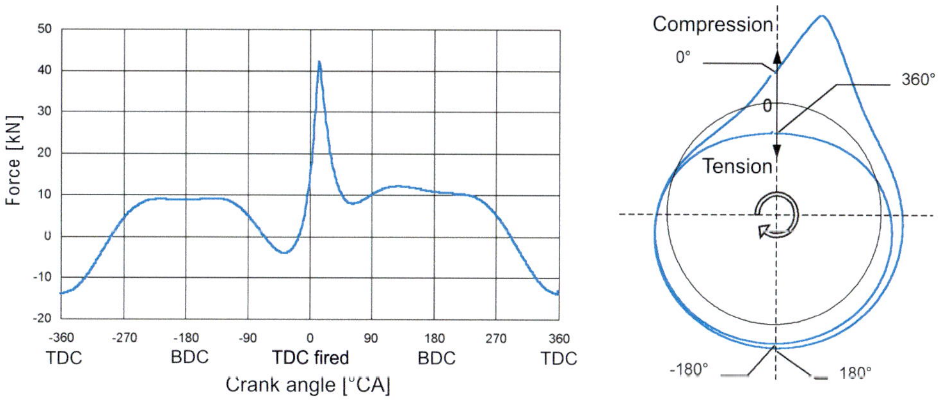


**Figure 2.1:** Piston pin load

For pistons in gasoline engines of passenger cars with moderate specific power output, the piston pins can be fixed in the small end bore with shrinkage stresses (fixed-pin connecting rod) (**Figure 2.2d**). This design allows savings due to the elimination of the piston pin circlips and the bushing in the small end bore and makes automatic assembly of the piston, piston pin, and connecting rod easier for large-scale production of engines.

In highly stressed gasoline engines and in diesel engines, the piston pin "floats" in the small end bore (**Figure 2.2a–c**). It needs to be secured with piston pin circlips against sideways motion in the piston (see Section 2.8).

In large-bore pistons, the cooling oil is often fed through the connecting rod and the piston pin, which features special oil feeding systems, to the piston pin boss; see **Figures 2.12–2.15**.

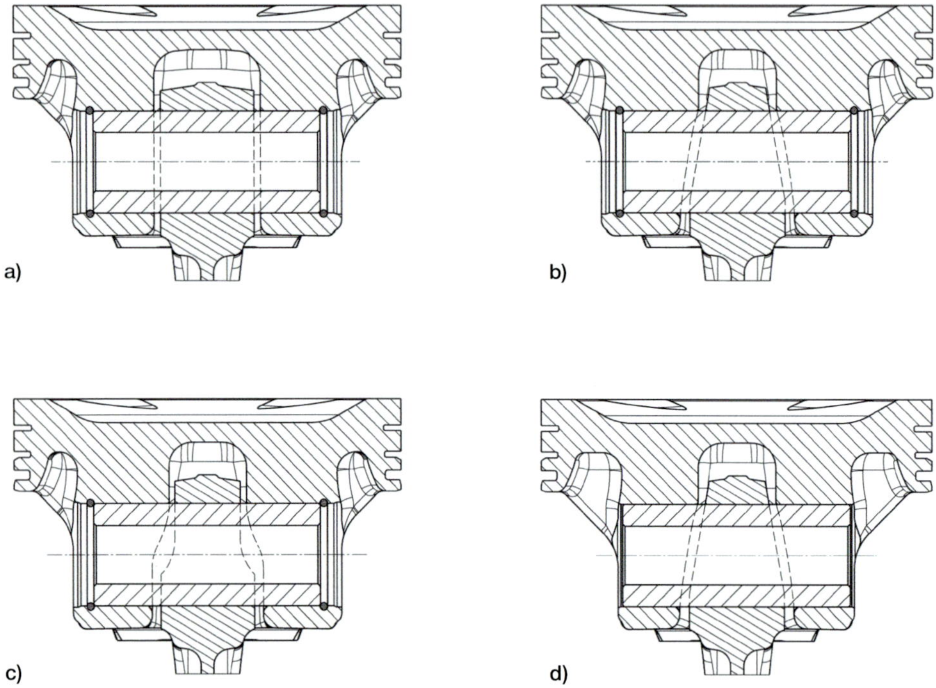


**Figure 2.2:** a) floating configuration with parallel support, b) floating configuration with tapered support, c) floating configuration with stepped support, d) fixed-pin connecting rod

# 2.2 Requirements

## 2.2.1 General

Piston pins must meet the following requirements:

- Sufficient strength and toughness to withstand the loads without damage
- High surface hardness, in order to achieve favorable wear behavior
- High surface quality and shape accuracy for optimal fit with its sliding partners, the piston and connecting rod
- Low weight, in order to keep inertia forces minimal
- Stiffness must be matched to the piston design, in order to avoid overloading the piston.

Despite these sometimes contradictory requirements, piston pin manufacture must be as simple, and thus economical, as possible.

## 2.2.2 Strength

Under the effects of the gas and inertia forces, pressure and stress loads act on the piston pin surface, the distribution of which is determined by the deformations of the piston pin bores, piston pins, and small end bore, caused by the forces (see Section 2.4.3). As a result of this pressure distribution, the piston pin is subjected to bending, ovalization, and shearing off. Added to this is a torsional load due to the connecting rod tilting motion. It is neglected because of its limited proportion in the total load. The opposing requirement is that the piston pin must be as stiff and as light as possible.

**Figure 2.3** shows the stress distribution on the piston pin during ovalization and various microstructure states at the surface.

The ovalization of the piston pin results in the stress distribution shown in **Figure 2.3a**. The maximum tensile stresses critical for fatigue resistance are inside, on the surface of the bore. Residual compressive stresses applied at the inner bore can counteract these tensile stresses, which has a positive effect on the fatigue resistance of the piston pin. The same applies analogously for the outer diameter, which is loaded mainly through bending.

The carbon and nitrogen uptake in the surface layer, associated with case hardening or nitriding of the piston pin, results in an increase in volume and thus residual compressive stresses in the layer. The effect on the residual stress state of the piston pin is shown in **Figures 2.3b–d**. Practical experience confirms that this significantly increases fatigue resistance.

Decarburization of the skin of the bore surface (**Figure 2.4**), which leads to residual tensile stresses (**Figure 2.3d**), is extremely detrimental to the fatigue resistance of the piston

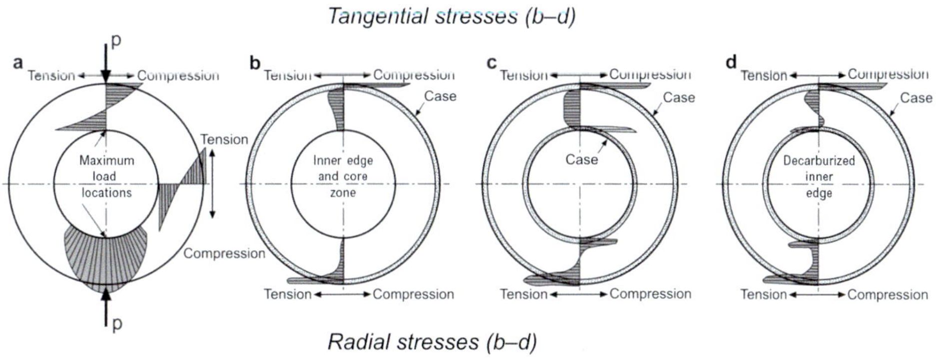


**Figure 2.3:** Stress distribution on the piston pin
a) effect of ovalization, b) without case hardening at inner bore, c) with case hardening at inner bore, d) during decarburization at inner bore

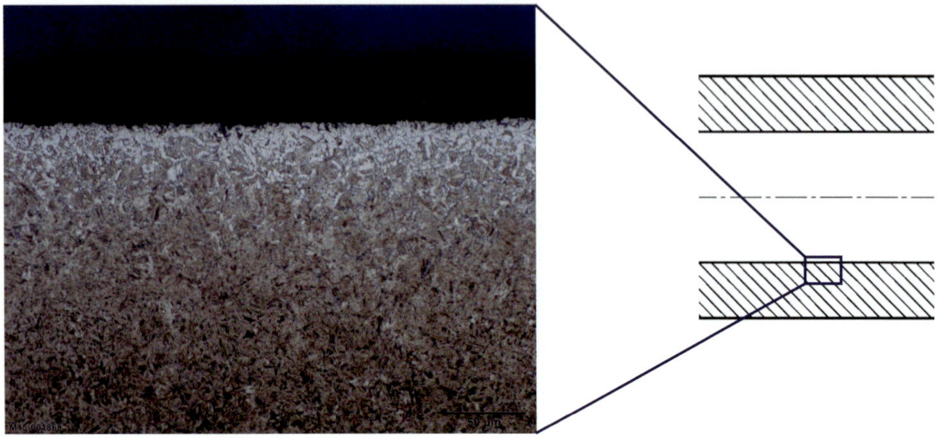

**Figure 2.4:** Decarburization of the surface at the bore of the piston pin

pin. Hardening cracks, slag lines, and deep machining lines in the bore also greatly reduce fatigue resistance.

Floating piston pins can turn. This means that highly loaded positions of the piston pin can move into less highly loaded positions, or form tensile to compressive loads, and vice versa. This results in a varying load on the piston pin. These stress amplitudes result in higher loading of the component, in contrast to piston pins that are fixed in the connecting rod, and therefore do not rotate. **Figure 2.5** shows the differences between a fixed and a rotating piston pin, using stress amplitudes.

The pin loads are evaluated using a fatigue strength map, e.g., according to Smith. Such a fatigue strength map must be determined for each material in use. Its limit lines correspond to the safety factor S = 1. The permissible minimum safety factor is determined according to the requirements and expected loads for each area of application, such as passenger cars, commercial vehicles, or motorsport.

Clearance between the piston pin and the piston pin boss or small end bore should be selected such that scuffing cannot occur between the contact points with the piston and the connecting rod. The clearance should be checked carefully, especially under warm operating conditions, because of the different thermal expansion coefficients of the materials used.

In order to avoid pin boss cracks, limits of the temperature-dependent material and load factors, such as contact pressure in the boss, must not be exceeded.

## 2.2.3 Deformation

Another requirement is that the piston pin must be light, in addition to having sufficient stiffness and strength. Stiffness relative to bending can be increased greatly, as the fourth power

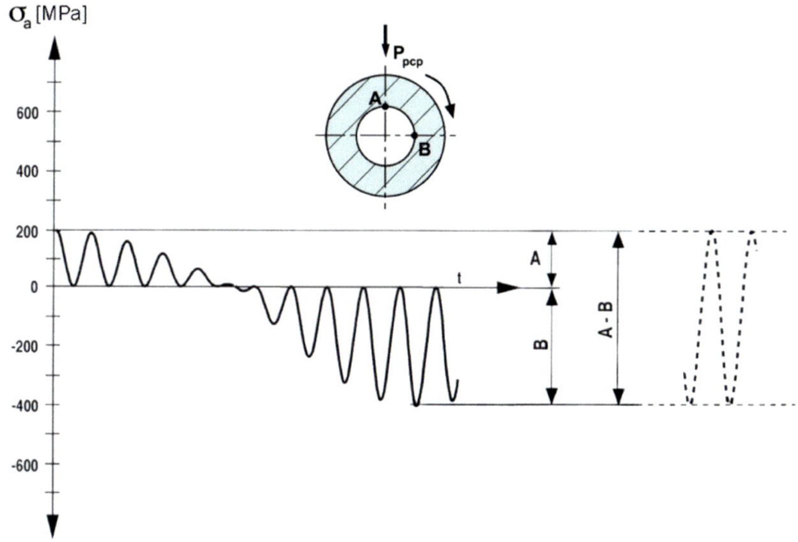


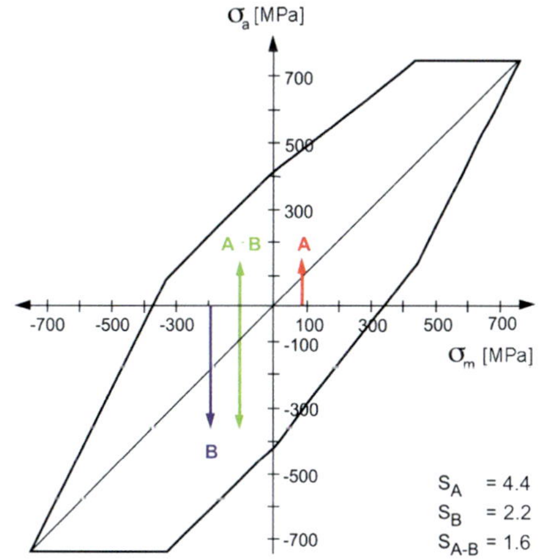


**Figure 2.5:**
Stress in a piston pin fixed in a connecting rod (A, B) and a rotating piston pin (A-B)

of the increase in diameter. Deflection also increases approximately as the third power of the support span of the piston pin, i.e., with the piston pin boss spacing. A reduction in this value thus causes a severe reduction in bending and thus increases stiffness. If a shorter piston pin can be used, then mass reduction is also possible.

An increase in stiffness relative to ovalization can be achieved only with a greater wall thickness and thus always increases mass. The stiffness of the piston pin has a significant effect on the loads on the piston pin boss, support, and bowl rim, as shown in **Figure 2.6**.

The susceptibility of the piston to pin boss cracks is shown in **Figure 2.7** as a function of the piston pin geometry, as a result of engine testing. Owing to higher peak cylinder pressures,

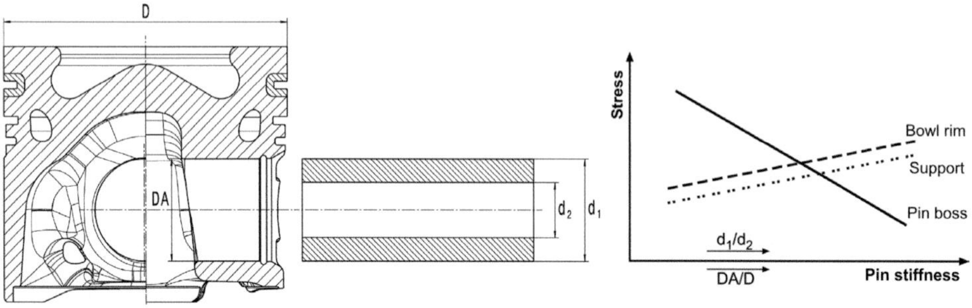


**Figure 2.6:** Piston stress as a function of piston pin stiffness

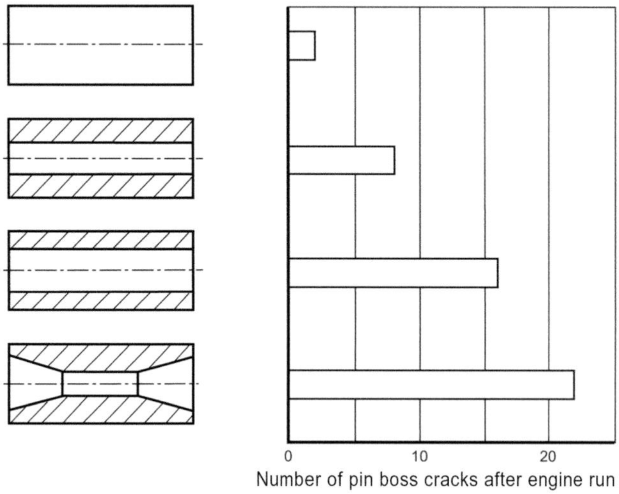


**Figure 2.7:** Boss stiffness as a function of piston pin geometry

diesel engines require stiffer piston pins in comparison to gasoline engines. The limit of maximum allowable contact pressure in the piston pin boss also demands larger pin diameters. Nevertheless, because of greater peak cylinder pressures in turbocharged engines, for example, piston pin bosses can be overloaded.

If potential piston design measures for reducing the critical stresses in the area of the piston pin boss have been exhausted, such as by increasing the piston pin outer diameter, reducing the pin boss spacing, and so forth, then a solution can be found with the use of shaped pin bores in the piston pin boss or profiled pins (**Figure 2.11**). These significantly reduce the stresses in the piston pin boss by means of a softer fit between the piston pin and boss. The diameter of the pin bore is slightly retracted in the area of the inner or top edges, according to the load. A smooth transition must be ensured.

### 2.2.4 Lubrication, oil supply

The sliding partners are mechanically loaded by gas and inertia forces. The transient loads cause alternating pressure on the bearing surfaces, such that boundary lubrication conditions can occur. The splash oil in the crankcase is not always sufficient to keep wear at a low level. The buildup of a lubricating film must then be supported by design measures. In the small end bore, this is carried out–in the case of large pistons–with splash oil feeders or pressurized oil supply through the connecting rod. Oil pockets can also be used as a reservoir. Pockets, oil grooves, and the like are incorporated in the piston pin boss.

### 2.2.5 Wear

Boundary lubrication conditions cannot be avoided under all operating conditions. Therefore, the contact between the piston pin and the small end bore and the piston pin boss bore must also have sufficient boundary lubrication properties and be wear-resistant. Given a high surface quality and hardness on the piston pin, this can be achieved in a simple manner. Piston pins are therefore case hardened or nitrided.

In the case of particularly high requirements for the surface, such as in motorsport, or if a bushingless connecting rod is used, the sliding properties (friction, wear resistance) can be significantly improved by an additional PVD or DLC coating (physical vapor deposition, PVD; diamond-like carbon, DLC). Coatings of this type allow ultrahigh contact pressures and reduce friction.

### 2.2.6 Weight

The total oscillating mass can be reduced by reducing the piston pin mass. The proportion of oscillating mass made up by the piston pin can be between 10 and 30%.

## 2.3 Types of piston pins

In most applications, the tubular or cylindrical piston pin (**Figure 2.8**) has been accepted as the standard design. It optimally fulfills requirements with regard to simple geometry and economical manufacture.

In order to reduce the inertia forces of drive unit components moving back and forth (oscillating), the mechanically less loaded ends of the pin bore can be designed with a conical shape to save weight (**Figure 2.9**).

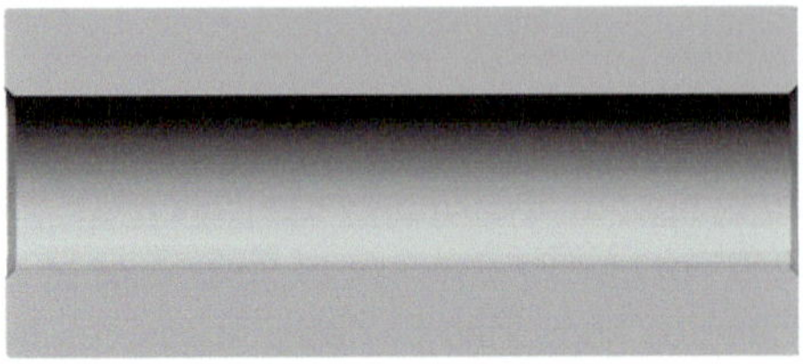

**Figure 2.8:** Piston pin with cylindrical bore

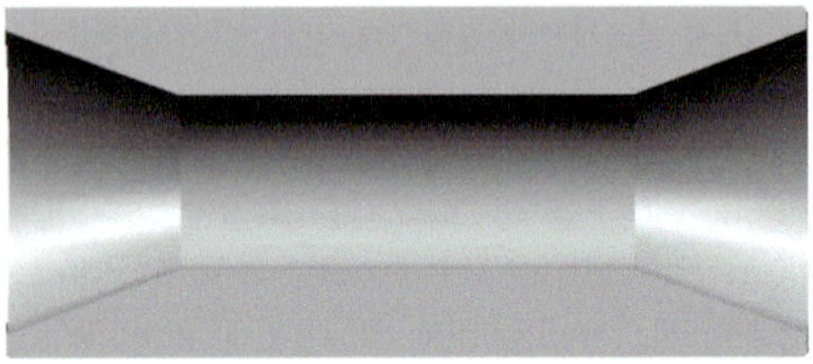

**Figure 2.9:** Piston pin with inner cones

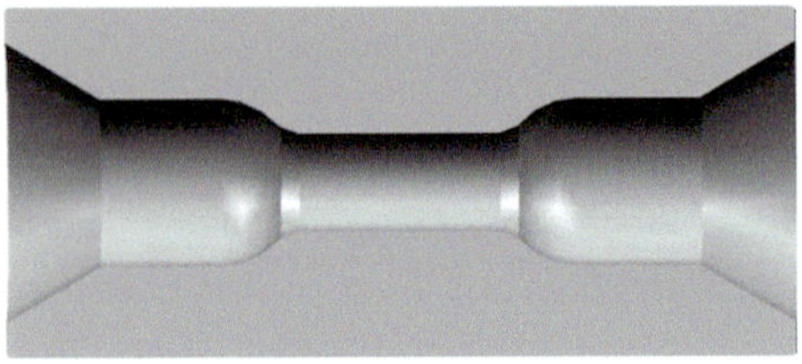

**Figure 2.10:** Piston pin with profiled inner contour

**Figure 2.11:** Piston pin with outer contour (profiled piston pin)

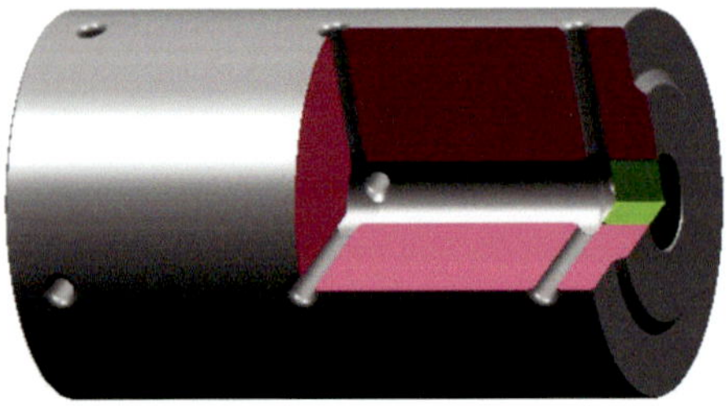

**Figure 2.12:** Piston pin with oil bores and blanking plugs (shrink-fit)

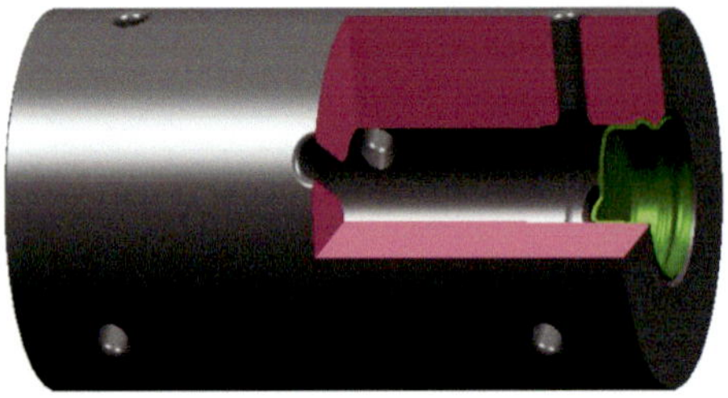

**Figure 2.13:** Piston pin with oil bores and sealing cover (rolled-in)

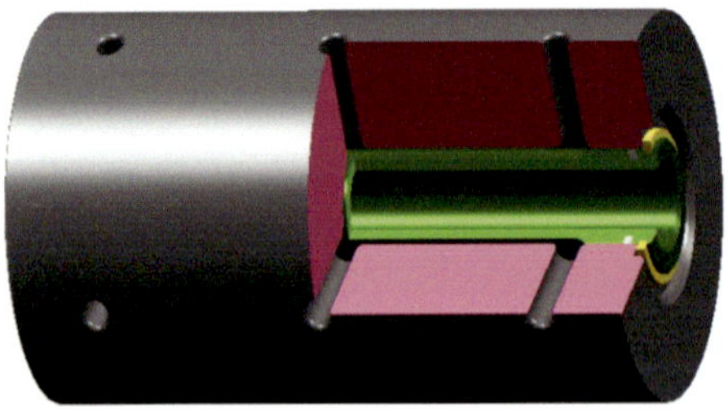

**Figure 2.14**: Piston pin with oil bores and oil feeding tube

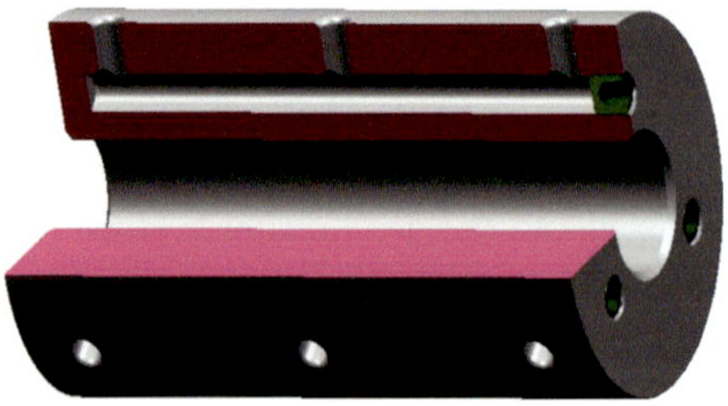

**Figure 2.15:** Piston pin with oil bores and screw plugs

Another piston pin variant, used especially for highly loaded diesel engines, is the inner contour piston pin (**Figure 2.10**). The wall thickness of the piston pin is reinforced specifically in the connecting rod area, while the ends of the piston pin contribute to mass reduction with a conical design.

For critical stresses in the piston pin boss and if the design options for the piston have been exhausted, the piston pin with a profiled outer contour can provide a solution (**Figure 2.11**). The outer surfaces of these piston pins are slightly retracted (approx. 20 to 40 μm) by profile grinding in the area of contact of the inner bore edges of the piston pin boss. It is crucial that the transitions from the undercut to the cylindrical areas are smooth and gradual.

For cooled pistons, especially large-bore pistons, the cooling oil is often fed from the connecting rod to the piston via the piston pin. Piston pins for oil-cooled pistons allow various design options (**Figures 2.12–2.15**). Secure closure of the piston pin on the face side under all conditions is of decisive importance for the cooling oil supply to the piston, and thus for the operational safety of the engine. Both during manufacture and in later operation, the piston pin with a shrink-fit plug has proven itself especially well (**Figure 2.12**).

## 2.4 Design

### 2.4.1 Dimensioning

Piston pins are designed for loading by gas and inertia forces, contact pressure, and deformation. The bearing clearance between the piston pin and the piston pin boss and small end bore must also be determined, in order to ensure trouble-free operation, that is, quiet piston action and minimal wear. Consideration must be given to the fact that when the thermal expansion of the piston–piston pin–connecting rod system varies, the clearance can be larger than the installation clearances for a warm engine and smaller at cold temperatures. The temperature dependence of the bearing clearance between the piston pin and small end bore is generally disregarded.

When designing the smallest relative bearing clearance in aluminum pistons (**Table 2.1**) in gasoline engines, differentiation must be made between a "floating" pin bearing and a piston pin with a shrink fit in the small end bore. A piston pin with a floating configuration is the standard design and is the variant that can be loaded the most specifically in the piston pin boss. With the shrunk connecting rod design, the piston pin is seated in the small end bore with some overlap.

Advantages and disadvantages of fixed-pin connecting rods and floating configuration of the piston pin in the connecting rod are shown in **Table 4.2**.

**Table 2.1:** Smallest relative installation clearance between the piston pin and piston or connecting rod for gasoline and diesel engines, motorsport engines not included

| Application | | Piston material | Piston pin bearing | Relative bearing clearance[1] | |
|---|---|---|---|---|---|
| | | | | Piston pin boss | Conrod bore |
| Gasoline engines | Pass. car | Al | With shrink fit connecting rod | > 0.4 ‰ | < −1.0 ‰ (overlap) |
| | Pass. car | Al | Floating | > 0.2 ‰ | > 0.4 ‰ |
| Diesel engines | Pass. car | Al | Floating | > 0.2 ‰ | > 0.6 ‰ |
| | Com. veh. | Al | Floating | > 0.2 ‰ | > 1.0 ‰ |
| | Com. veh. | St | Floating | > 1.0 ‰ | > 1.0 ‰ |
| Large-bore engines | | St/Al | Floating | > 0.15 ‰ | > 1.0 ‰ |
| | | St/St | Floating | > 0.5 ‰ | > 1.0 ‰ |

[1] relative to the outer diameter of the piston pin

The piston and connecting rod geometry and the maximum pressure in the expansion stroke cycle must be considered when dimensioning the piston pin. Depending on the application, dimensions according to **Table 2.2** are the result.

**Table 2.2:** Typical major dimensions of piston pins
D: piston diameter, $d_1$: piston pin outside diameter, $d_2$: piston pin inside diameter, l: piston pin length

| Application | | Piston | Piston pin | | |
|---|---|---|---|---|---|
| | | D [mm] | $d_1/D$ | $d_2/d_1$ | l/D |
| Gasoline engines | 2-stroke | 35–70 | 0.20–0.30 | 0.40–0.73 | 0.65–0.80 |
| | Pass. car | 65–100 | 0.20–0.30 | 0.47–0.60 | 0.60–0.75 |
| Diesel engines | Pass. car | 65–95 | 0.30–0.40 | 0.43–0.53 | 0.65–0.80 |
| | Com. veh. Al | 100–160 | 0.40–0.45 | 0.40–0.47 | 0.78–0.82 |
| | Com. veh. St | | | 0.31–0.47 | 0.60–0.85 |
| Large-bore engines | | < 250 | 0.30–0.45 | 0.34–0.56 | 0.70–0.86 |
| | | > 250 | 0.35–0.45 | 0.38–0.45 | 0.65–0.86 |

## 2.4.2 Analysis

An analysis of the transient deformations and stresses on the piston pin cannot be performed very accurately, even with great effort, because the following factors, amongst others, need to be considered simultaneously:

- Significantly different piston cross sections, and thus stiffnesses, required for functional purposes
- Effect of the piston temperature on piston deformations and on piston stiffness (Young's modulus)
- Effects of piston pin deformation
- Different Young's modulus of the piston material and piston pin material
- Different elastic section moduli of piston pin cross sections (e.g., conical piston pins)
- Lubricating film distribution

Using simplified load assumptions, analyses can be performed that, together with empirical values, enable an assessment of the operating conditions. Assuming a surface load in the conrod bore and individual point loads in the pin bores in the piston, Schlaefke presented a useful calculation method back in 1940 (**Figure 2.16**).

In addition to the deformation due to bending and ovalization, the "total stress" is determined from the bending stress $\sigma_B$ and the stress due to ovalization $\sigma_A$.

It is assessed on the basis of empirical values for total stress and deformation. The average pin bore pressure must not exceed the threshold prescribed by the piston strength.

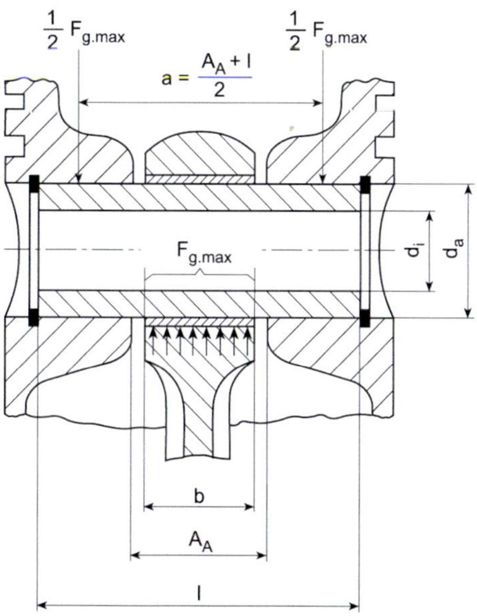


Stress due to ovalization

$$\sigma_A = \frac{3F_{g,max}(d_a + d_i)}{4l(d_a - d_i)^2}$$

Stress due to bending

$$\sigma_B = \frac{8F_{g,max}\, a\, d_a}{\pi(d_a^4 - d_i^4)}$$

Total stress

$$\sigma_{ges} = \sqrt{\sigma_A^2 + \sigma_B^2}$$

**Figure 2.16:**
Load schematic of a piston pin (Schlaefke design)

## 2.4.3 Finite element analysis

As for other components, the use of finite element analysis methods (FE) in component design has also been accepted for piston pins. The EHD contact (elasto-hydrodynamic contact) must be calculated under consideration of the deformations and lubricant gap geometry. This analysis is very computation-intensive, since the deformations due to temperature, gas, and inertia force loads on the piston and connecting rod need to be considered.

Boundary conditions of the EHD contact at the piston pin, defined by the load case, have been standardized for variant analyses and a simplified 3D FE calculation method has been derived.

The MAHLE program MPOT uses a pressure distribution in the connecting rod and the piston pin boss for load introduction. This pressure distribution has been determined for pistons using a 3D FE analysis and is the basis of the program as a standardized elasto-hydrodynamic lubrication pressure distribution. Pressure profiles have been calculated and

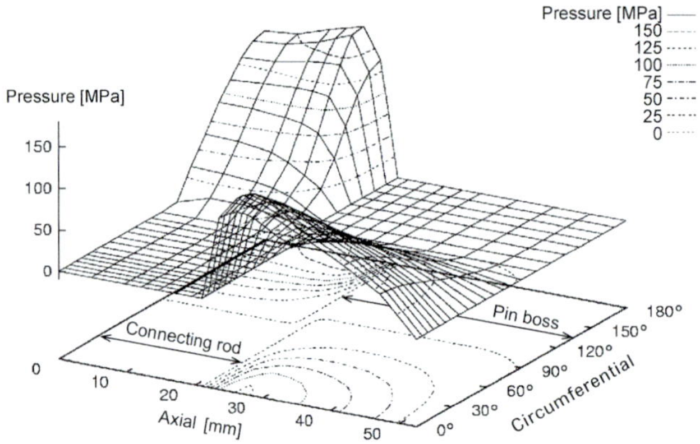


**Figure 2.17:**
Pressure distribution for parallel support of a piston pin

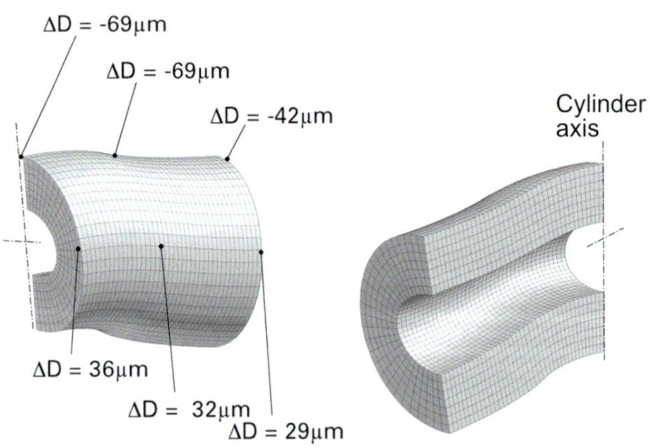


**Figure 2.18:**
Deformation of a piston pin (large-bore engine) analyzed with MPOT

integrated for all applicable support cases. **Figure 2.17** shows an example of a pressure distribution for parallel support. With the aid of the peak cylinder pressure and the geometric data (piston diameter, boss, piston pin, and connecting rod geometries), the corresponding profile is applied to the new data and a mesh for a quarter of a piston pin is generated automatically. The results are available after just a few minutes of computation (see **Figures 2.18–2.20**).

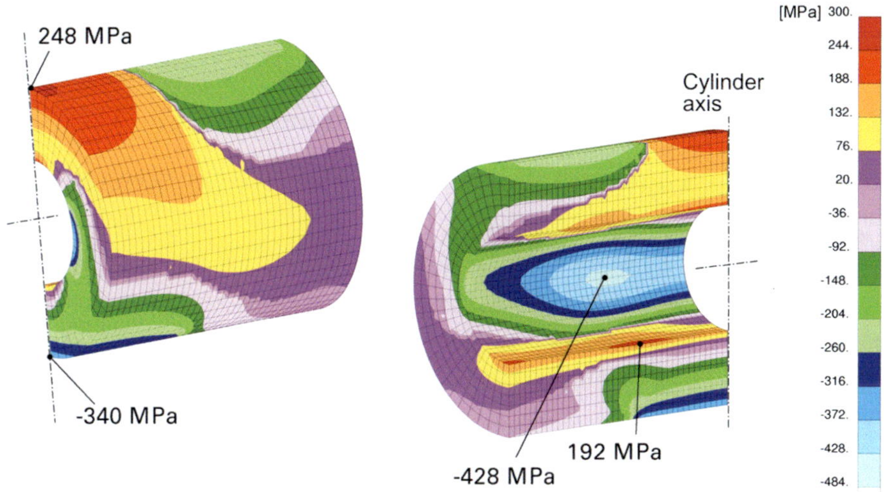


**Figure 2.19:** Analysis of main stresses on the piston pin (large-bore engine)

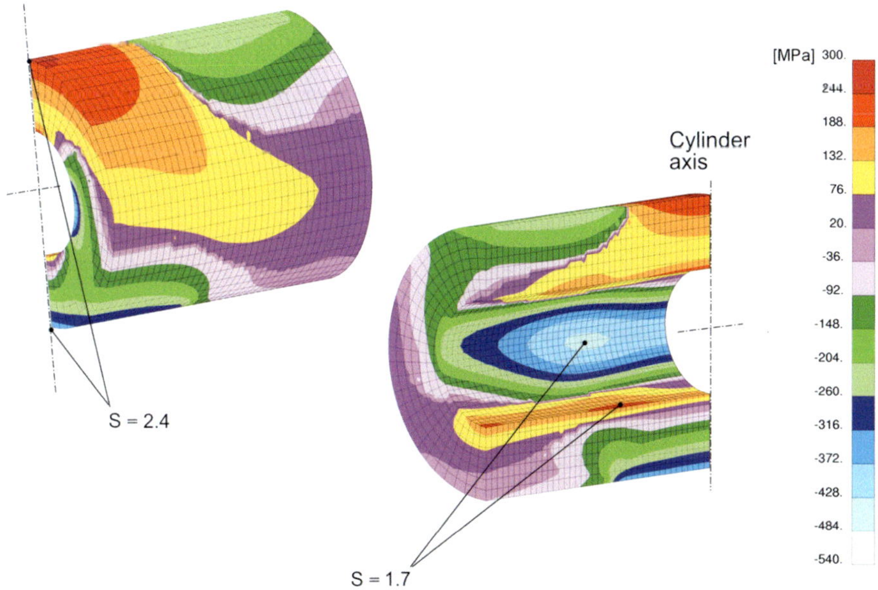


**Figure 2.20:** Safety factors at various locations of the piston pin (large-bore engines)

The MAHLE program MPOT enables the simplified design of piston pins for passenger car and commercial vehicle aluminum pistons with cylindrical piston pin shapes and tapered bores. Parallel, keystone, and stepped support geometries are available.

Assessment of the calculated stresses (**Figure 2.19**) is carried out automatically, using the integrated accessory program, for typical piston pin materials, and safety factors are output (**Figure 2.20**).

### 2.4.4 Dimensional and form tolerances, standard

The markings on the piston pin corresponding to piston pin standard ISO 18669 are shown in **Figure 2.21**.

The piston pin standard DIN 73216 has been internationally revised and published as ISO 18669-1 and 18669-2. Part 1, “General Specifications,” lists the markings, piston pin types, dimensions and tolerances, materials, heat treatment, and quality characteristics. Part 2 deals with measurement and test methods.

MAHLE piston pins are designed, manufactured, and applied on the basis of the ISO 18669 standard.

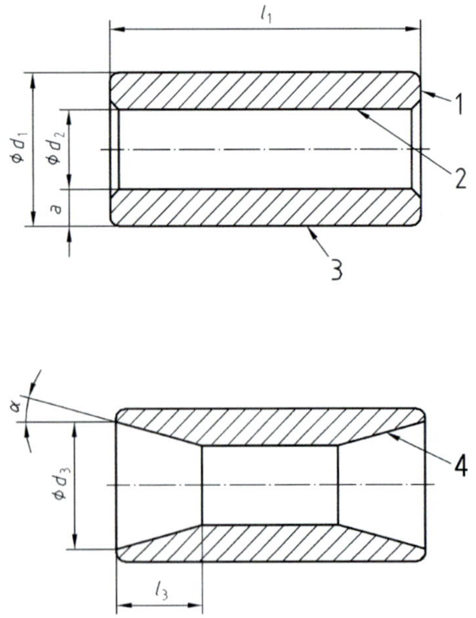


$d_1$: outer diameter
$d_2$: inner diameter
$l_1$: length
$a$: wall thickness
1: end surface
2: bore surface (inner surface)
3: outer surface

$d_3$: tapered outlet diameter
$l_3$: taper length
$\alpha$: taper angle
4: tapered bore surface

**Figure 2.21:** Markings on a piston pin

The important design criteria listed in the standard–core hardness, hardness penetration depth, surface hardness, volume stability, and surface roughness–are provided in **Tables 2.3–2.6**.

**Table 2.3:** Core hardness (core strength)

<table>
<tr><th rowspan="2">Wall thickness a [mm]</th><th colspan="3">Core hardness HV 30 (core strength $R_m$ [MPa]) [1)]</th></tr>
<tr><th>Class L</th><th>Class M</th><th>Class N</th></tr>
<tr><td>1.5–2</td><td>310–515<br>(1,000–1,650)</td><td rowspan="2">310–470<br>(1,000–1,500)</td><td rowspan="5">310–470<br>(1,000–1,500)</td></tr>
<tr><td>>2–5</td><td>280–485<br>(900–1,575)</td></tr>
<tr><td>>5–10</td><td>270–470<br>(850–1,500)</td><td rowspan="3">280–470<br>(900–1,500)</td></tr>
<tr><td>>10–15</td><td>250–470<br>(800–1,500)</td></tr>
<tr><td>>15–25</td><td rowspan="2">235–470<br>(750–1,500)</td></tr>
<tr><td>>25</td><td>250–435<br>(800–1,400)</td><td></td></tr>
</table>

[1)] The core strength values ($R_m$) are provided for reference only and are calculated from the core hardness HV with a factor of 3.2.

**Table 2.4:** Hardness penetration depths, dimension in mm

<table>
<tr><th rowspan="3">Wall thickness a [mm]</th><th colspan="5">Case depth</th><th colspan="2">Nitride depth</th></tr>
<tr><th colspan="2">Outside</th><th rowspan="2">Inside min.</th><th colspan="2">Outside and inside together</th><th rowspan="2">Outside min.</th><th rowspan="2">Inside min.</th></tr>
<tr><th>Min.</th><th>Code X</th><th>Max.</th><th>Code X</th></tr>
<tr><td>1.5–<2</td><td>–</td><td>0.4</td><td>0.1</td><td>0.65 · a</td><td>0.80 · a</td><td rowspan="5">0.3</td><td rowspan="5">0.2</td></tr>
<tr><td>2–3</td><td>0.3</td><td>0.5</td><td>0.1</td><td>0.65 · a</td><td>0.80 · a</td></tr>
<tr><td>>3–5</td><td>0.4</td><td>0.6</td><td>0.2</td><td>0.50 · a</td><td>0.65 · a</td></tr>
<tr><td>>5–15</td><td>0.6</td><td>–</td><td>0.4</td><td>0.35 · a</td><td>–</td></tr>
<tr><td>>15</td><td>0.8</td><td>–</td><td>0.6</td><td>0.35 · a</td><td>–</td></tr>
</table>

Comment 1: the limit hardness used in determining the case depth is Hs 550 HV.

Comment 2: for piston pins with limited change in volume, identification mark V, the limit hardness is Hs 500 HV.

Comment 3: code X: applies to piston pins used with needle bearing in the conrod bore.

**Table 2.5:** Surface hardness for Class 1 piston pins

| Hardness measuring method | Surface hardness | | |
|---|---|---|---|
| | Case-hardened steel | | Nitrided steel |
| | Unlimited change in volume | Limited change in volume, abbreviation: V | |
| Vickers HV 10 | 675 min. | 635 min. | 690 min. |
| Rockwell HRC [1)] | 59 min. | 57 min. | – |
| Rockwell HRA [2)] | 80.7 min. | 79.6 min. | – |

1) Case depth min. 0.7 mm

2) Case depth 0.4–0.9 mm

**Table 2.6:** Volume change after heat resistance test, dimensions in mm

| Test conditions | Outer diameter $d_1$ | Max. increase in dimension $\Delta d_1$ | | |
|---|---|---|---|---|
| | | Case-hardened steel | | Nitrided steel |
| | | Unlimited change in volume | Limited change in volume, abbreviation: V | |
| After 4 h at 180°C | ≤50 | +0.006 | 0 | 0 |
| | >50–≤60 | +0.008 | 0 | |
| | >60–100 | +0.012 | 0 | |
| After 4 h at 220°C | ≤50 | – | +0.006 | |
| | >50–≤ 60 | – | +0.008 | |
| | >60–100 | – | +0.012 | |

## 2.5 Materials

MAHLE piston pins are manufactured from high-quality case-hardened or nitrided steels. Case or nitride hardening yields good toughness in the core and high surface hardness with good wear behavior. Piston pins made of nitrided steel are especially noteworthy for their outstanding wear resistance. The enrichment of the edge zones with carbon or nitrogen causes an increase in volume, which leads to compressive stresses in the piston pin edge layers. As previously indicated, these residual compressive stresses at the surface have a positive effect on the fatigue resistance of the piston pin. Material or microstructure defects, such as decarburization of the skin, cementite network, missing case hardening of the inner bore, hardening and grinding cracks, or open slag lines are especially critical in these edge zones.

Piston pins made of case-hardened steel bear the problem of lack of volume stability, i.e., with increasing surface hardness (increased residual austenite content), the piston pin diameter will continually "grow" under heat load (**Table 2.6**).

**Table 2.7** shows the composition, physical properties, and areas of application of MAHLE piston pin materials.

**Table 2.7:** MAHLE piston pin materials

| Chemical composition by weight % | Case-hardened steels | | | Nitrided steel |
|---|---|---|---|---|
| | 17Cr3 | 16MnCr5 | 17CrNi6 | 31CrMoV9 |
| | SAE 5115 (Class L)[1] | (Class M)[1] | | (Class N)[1] |
| C | 0.13–0.20 | 0.14–0.19 | 0.14–0.20 | 0.27–0.34 |
| Si | 0.15–0.40 | 0.15–0.40 | 040 max. | 0.40 max. |
| Mn | 0.60–0.90 | 1.00–1.30 | 0.50–0.90 | 0.40–0.70 |
| P | ≤0.035 | ≤0.035 | ≤0.035 | ≤0.025 |
| S | ≤0.040 | ≤0.035 | ≤0.035 | ≤0.035 |
| Cr | 0.70–1.00 | 0.80–1.10 | 1.40–1.70 | 2.30–2.70 |
| Ni | | | 1.40–1.70 | |
| Mo | | | | 0.15–0.25 |
| V | | | | 0.10–0.20 |
| Young's modulus [MPa] | 210,000 | 210,000 | 210,000 | 214,000 |
| Thermal expanson[2] [$10^{-6}$ 1/K] 20–200°C | 13.1 | 13.1 | 12.8 | 13.0 |
| Thermal conductivity[2] $\lambda$ [W/m*K] | 36 | 36 | 37 | 39 |
| Density [g/cm$^3$] | 7.82 | 7.84 | 7.84 | 7.83 |
| Poisson ratio $\mu$ | 0.27 | 0.27 | 0.27 | 0.27 |
| Application | Gasoline and passenger car diesel engines | High-performance passenger car engines and commercial vehicle and medium-speed diesel engines | Large-bore engines | Highly loaded gasoline and diesel engines |

[1] conforms to ISO 18669-1

[2] determined using separately produced samples of the same hardness (approx. 300 HV)

For highly stressed racing and motorsport engines and for all large-bore piston pins, ESR (electro slag remelting) quality steels are used. The ESR steels are exceptional for their very high degree of purity, low sulfur content, and high uniformity in microstructure. **Figure 2.22** shows typical hardness curves over the piston pin cross section with associated microstructure at the outside, in the core, and at the bore, for case-hardened and nitrided piston pins.

**Figure 2.22:** Typical hardness curve and microstructure of piston pins, case-hardened and nitrided

## 2.6 Coating

Various amorphous DLC coatings that contain hydrogen (a-C:H) are used for low-friction and secure operation of piston pin bearing (**Table 2.8**). The coatings are built up in layers, and the layer hardness is adapted to the contact loads and materials making contact with the piston pin. Three types of layers are distinguished after layer buildup: single, dual, and triple layer (**Figure 2.23**). Total thicknesses are between 2 and 3.5 μm. The layers with high hardness values are used for sliding contact surfaces made of abrasive materials, such as the aluminum piston alloy, or for high contact pressures.

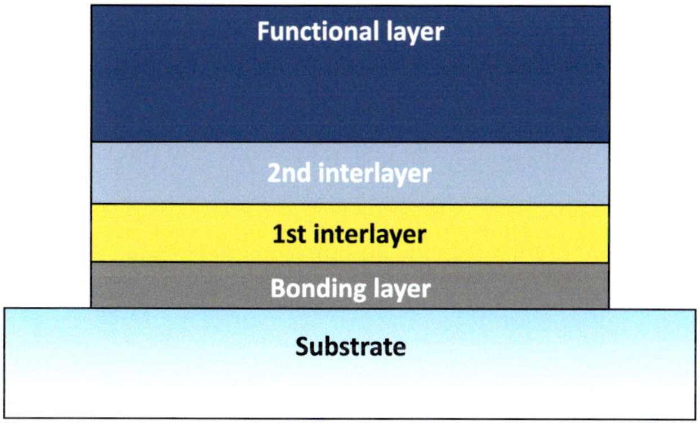


**Figure 2.23:** DLC layer buildup, example of a triple layer

**Table 2.8:** MAHLE piston pin coatings

| MAHLE piston pin coating | Type of layer | Layer buildup | Indentation hardness HIT [GPa] | Layer property |
|---|---|---|---|---|
| MPC-101 | Single layer | a-C:H | 20 | High wear resistance |
| MPC-102 | | | 24 | |
| MPC-201 | Dual layer | CrN, a-C:H | 20 | High layer strength |
| MPC-202 | | | 24 | |
| MPC-203 | | a-C:H:W, a-C:H | 20 | |
| MPC-204 | | | 24 | |
| MPC-301 | Triple layer | CrN, CrC, a-C:H | 20 | High wear resistance and very high layer strength |
| MPC-302 | | | 24 | |

## 2.7 Component testing

### Piston pin test bench

Piston pins are often tested on servo-hydraulic test machines and resonance pulsators. A simulation of the rotational motion of the piston pin is generally not included. As previously indicated, the loads on the piston pin in a floating configuration cannot be tested with sufficient accuracy using this method. Piston pins with a floating configuration are therefore tested on a special fixture, the piston pin test bench (**Figure 2.24**). With this test installation, the alternating loads on the rotating piston pin can be reproduced, with bending and ovalization.

The test load is applied statically and can be adjusted continuously up to the maximum load.

The piston pin is turned under load at a constant rpm. The rotational motion is transferred to the piston pin indirectly, without introducing a moment, by driving the boss bearing. The piston pin mount is a geometric reproduction of the real piston pin boss and the small end bore.

The piston pin load and deflection, bearing temperatures, and displacement of the connecting rod are all monitored. The system shuts down if the connecting rod changes position as a result of a crack in the piston pin.

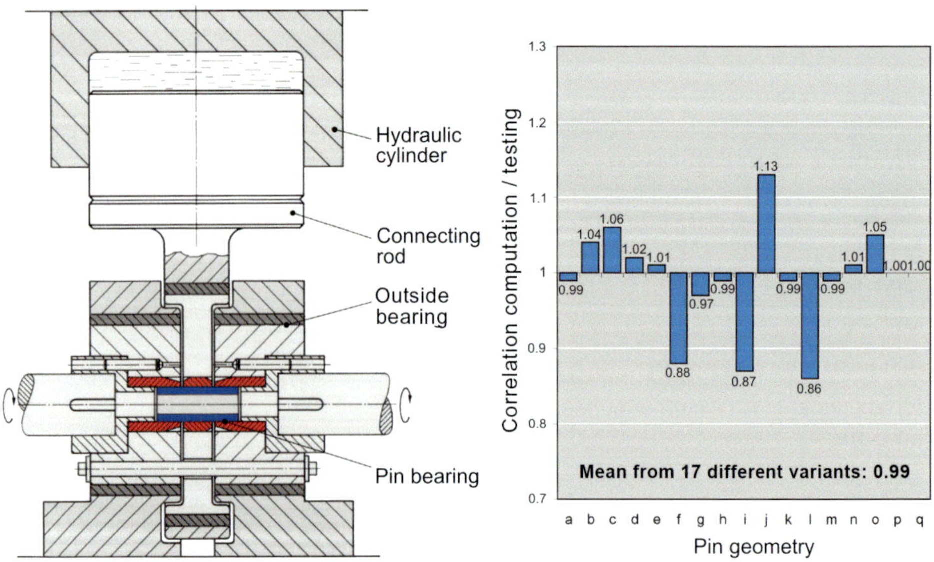


**Figure 2.24:** Passenger car piston pin test bench, correlation between analysis and testing

## 2.8 Piston pin circlips

If the piston pin is not held in the small end bore by a shrink-fit connection, then it must be secured to prevent it from moving sideways out of the piston pin boss and contacting the cylinder wall. For small and passenger car engines, this is solved almost exclusively with circlips mounted on the outside, made of round or square wire, which are inserted in corresponding grooves in the outside of the piston pin boss. Circlips made of round or square wire (also called snap rings) are made of patented drawn spring steel wire (DIN EN 10270-1) or oil-tempered spring steel wire (DIN EN 10270-2). **Figure 2.25** shows a typical round wire snap ring, such as is used in passenger car engines.

For easier assembly, the ends of the snap rings can be drawn in to form hooks (**Figure 2.26**).

The hooks, however, increase the mass at the ends of the rings and thus lead to lower engine speed to the point where the snap rings are lifted out of the circlip groove in the piston. Owing to this lower speed limit for snap rings with hooks, these circlips are used almost exclusively in diesel engines.

For high-speed engines, the seat of the circlip ends can be fixed in the groove by a hook that is bent outward, so that the joint opening is oriented in the direction of the stroke and the ring cannot rotate in the groove. The example in **Figure 2.27** shows the type and location of the ring gap, suitable for very high speed limits.

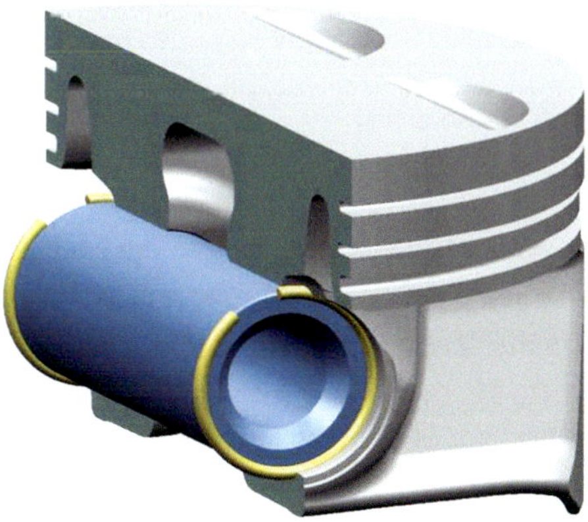

**Figure 2.25:**
Pistons for passenger cars with round wire snap ring, shape C, per DIN 73130

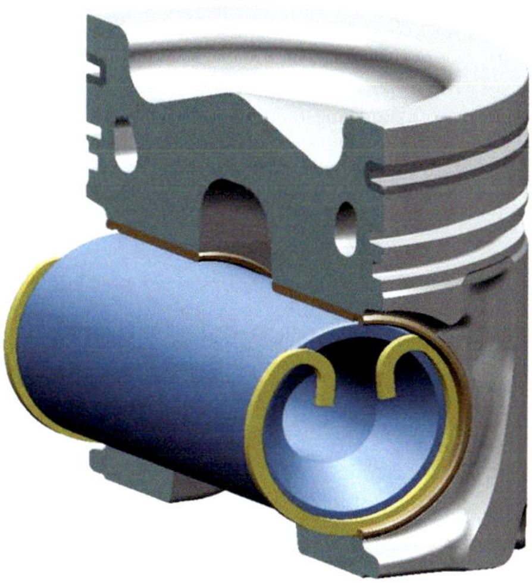

**Figure 2.26:**
Diesel engine piston with pin bore bushing and square wire snap ring

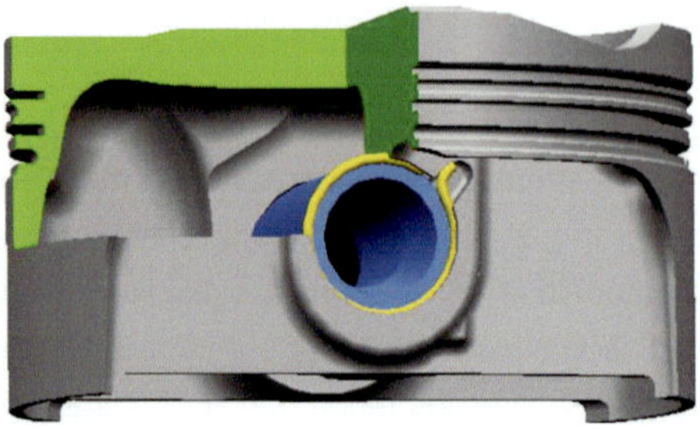

**Figure 2.27:**
Snap ring with external hook for very high speed limits

For large piston pin diameters, eccentrically stamped circlips according to DIN 472 and, increasingly, rings made of square wire with hooks are employed. See **Figures 2.28a–c**. So-called oval snap rings are used in connection with large-bore pistons with long piston pins.

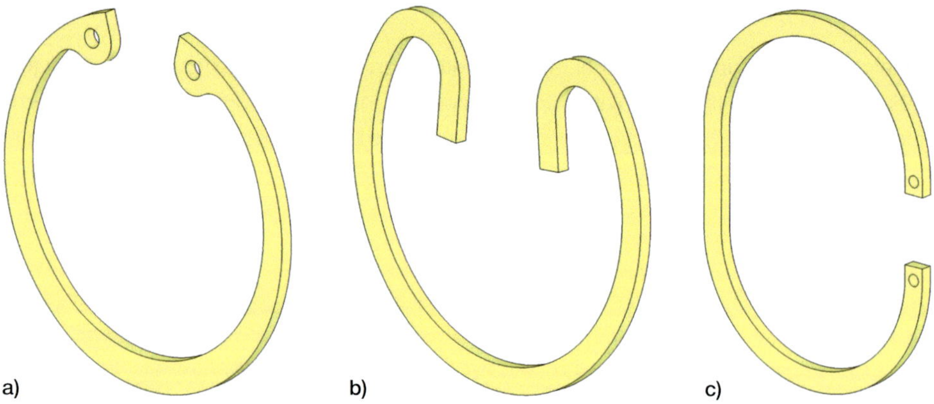


**Figure 2.28:** Circlips for large piston pins
a) DIN 472 seeger circlip ring, b) square wire snap ring, c) oval snap ring

Internal tension circlips per DIN 471 can also be used. These locking devices are installed in grooves at the end of the piston pin. The piston pin must then be longer and therefore heavier in comparison with a design that uses external tension circlips. No circlip groove is needed in the pin bore on the piston side. Producing the groove in the piston pin is difficult and is associated with higher costs, which is why this form of retaining the piston pin is used only very rarely.

# 3 Bearings

## 3.1 Product range

Bearings are used to ensure the function of the movable connection between two components. In general, different types of bearings include roller, plain, air, liquid, and magnetic bearings. The MAHLE product range focuses on bearing shapes for engines and peripheral applications.

### 3.1.1 Applications

Bearings are needed to locally separate surfaces that move relative to each other. This is achieved by a viscous lubricating film generating a pressure field that withstands even very high external loads, if the surfaces and their relative motion are properly designed.

Most bearings in the MAHLE product range are used in the engines of motor vehicles:

- Connecting rod bearing for the big end bore (radial)
- Main bearing (radial)
- Flange bearing (radial and axial)
- Thrust washers (axial)
- Conrod bushings for the small end bore (radial)

Other applications for MAHLE bearings are

- bushings for camshafts;
- bushings and washers for other automotive systems, such as transmissions.

**Figure 3.1** shows the variety of bearings that are installed in an engine.

### 3.1.2 Types and terminology

A distinction is generally made between bimetal and trimetal bearings. Bimetal bearings include radial plain bearings, bushings, and thrust washers. They generally consist of a steel support shell with an aluminum or bronze alloy coating. Trimetal bearings consist of a steel support shell coated with an aluminum or bronze alloy, with a thin bearing layer (galvanic, polymer, or sputter layer) referred to as an overlay. Typical bearing designs and terms are shown in **Figures 3.2–3.7**.

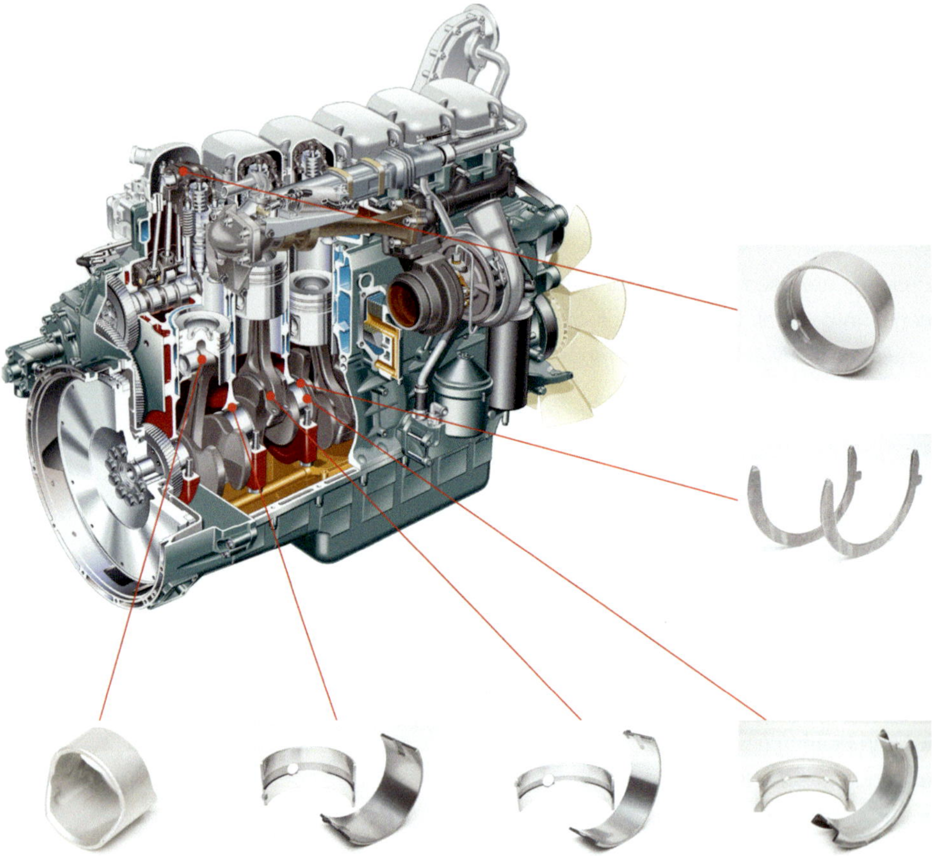

Figure 3.1: Plain bearing applications in a combustion engine

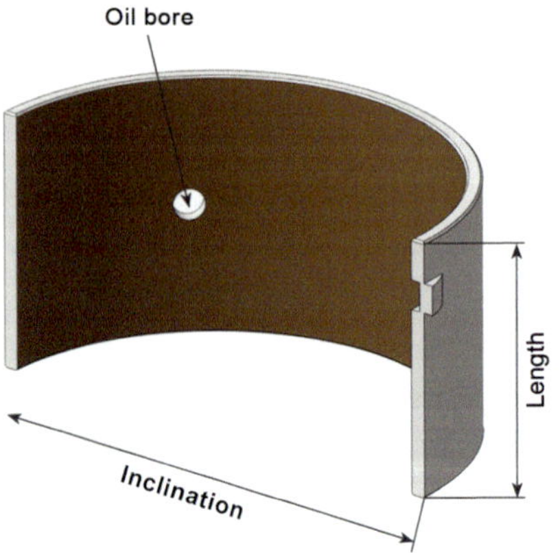


**Figure 3.2:** Conrod bearing

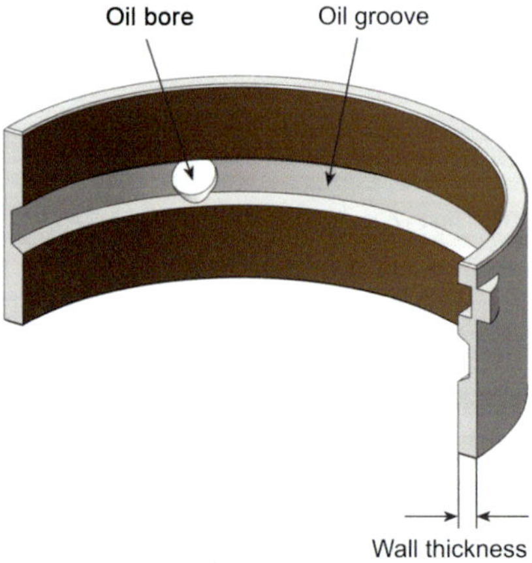


**Figure 3.3:** Main bearing shell (crankshaft)

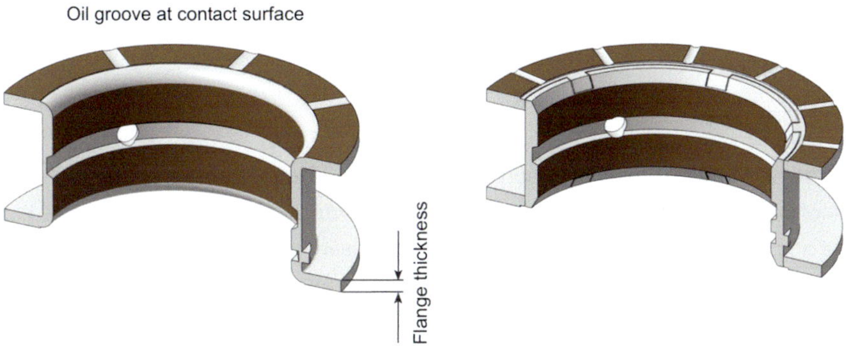


**Figure 3.4:** Flange bearing–solid bearing (rigid) and trimetal bearing (flexible)

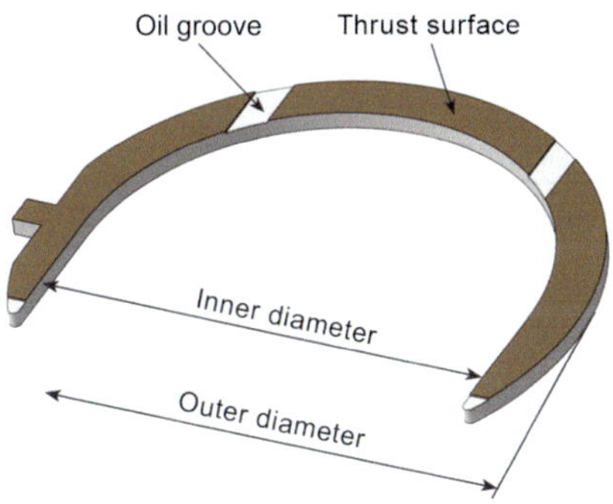


**Figure 3.5:**
Thrust washer for axial bearing

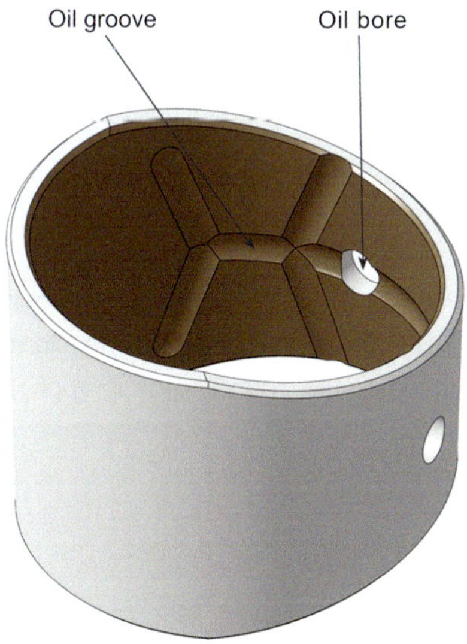


**Figure 3.6:** Connecting rod bushing

**Figure 3.7:** Camshaft bushing

# 3.2 Design specifications

## 3.2.1 Properties

A prerequisite for correct material selection, relative to the application profile of the engine, is knowledge of the material properties. The bearing loads occurring in the engine describe the mechanical and tribological requirements profile for the bearing. Material selection is always the result of a compromise among all the properties, which are often contradictory in nature. Important definitions and properties are explained in **Table 3.1**.

**Table 3.1:** Important bearing properties

| Property | Description |
|---|---|
| Load carrying capacity | Ability to bear mechanical loads on a sustained basis |
| Wear resistance | Resistance of the material to sliding wear |
| Seizure resistance | Ability of the material to run at the lubrication limit without welding to the journal; it depends on whether soft phases are present in the material composition |
| Embeddability | Ability of the material to tolerate and absorb hard particles on the sliding surface |
| Conformability | The ability to compensate for geometric deviations that cause local contacts |
| Corrosion resistance | Ability to resist corrosion by organic and mineral acids from combustion and oxidation of lubricants |

The most important properties are evaluated for each material (Section 3.5) and are used as an aid in material selection.

## 3.2.2 Load carrying capacity

Load carrying capacity is determined using what is known as a "Sapphire" test bench (**Figure 3.8**). The test bench consists of a motor-driven eccentric test shaft that exerts a load impulse on the test bearing shell with every revolution, while the counterforce is applied by a hydraulic cylinder. The test is run under lubricated conditions and the temperature is controlled. Test conditions are listed in **Table 3.2**.

Potential tests include

- screening test–brief repeated test under fixed load;
- L/N test–assessment of load cycles until damage under constant load;
- "staircase" test–a statistical assessment of load carrying capacity with stepwise increases.

The loads applied depend on the bearing shell length that is standardized for this test:

- 19.5 mm length – 70 to 180 MPa
- 29.5 mm length – 50 to 130 MPa

**Table 3.2:** Test conditions for the "Sapphire" load carrying capacity test

| **Test conditions** | |
|---|---|
| Skirt material | Hardened steel |
| Skirt speed | 3,000 rpm |
| Load | ~ 50 to 180 MPa |
| Lubricant | Synthetic 46 |
| Temperature | 110 °C |
| Test duration | ~ 3.6 million cycles |

**Figure 3.8:** "Sapphire" load carrying capacity test bench

### 3.2.3 Wear resistance

Wear resistance is evaluated using a test bench known as "Viper" (**Figure 3.9**). The test bench consists of an eccentric shaft against which the test bearing is pressed. The lever force is produced by the weight of a ballast and is transferred via a lever. Lubrication is applied continuously via a nozzle. The loss of mass is determined and then computationally converted into a loss of volume. The test conditions are summarized in **Table 3.3**.

**Tabelle 3.3:** Test conditions for the Viper wear resistance test

| **Test conditions** | |
|---|---|
| Shaft material | Hardened steel |
| Shaft speed | 500 rpm |
| Load | 0.04 MPa |
| Lubricant | Synthetic 46 |
| Temperature | 120 °C |
| Test duration | 60 minutes |

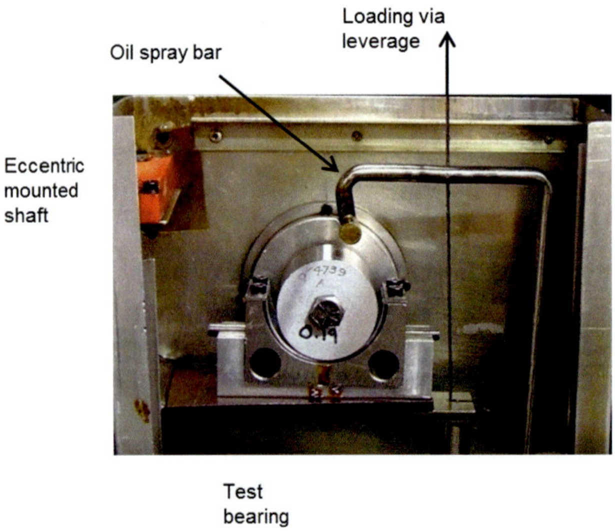


**Figure 3.9:** "Viper" wear resistance test bench

### 3.2.4 Stop-start applications

In order to evaluate load carrying capacity and wear resistance under dynamic lubrication conditions, particularly as they arise in conjunction with stop-start applications, a "Sapphire" test bench has been expanded to allow the load and speed to be controlled by a computer (**Figure 3.10**). The test bench motor has been programmed so that it per-

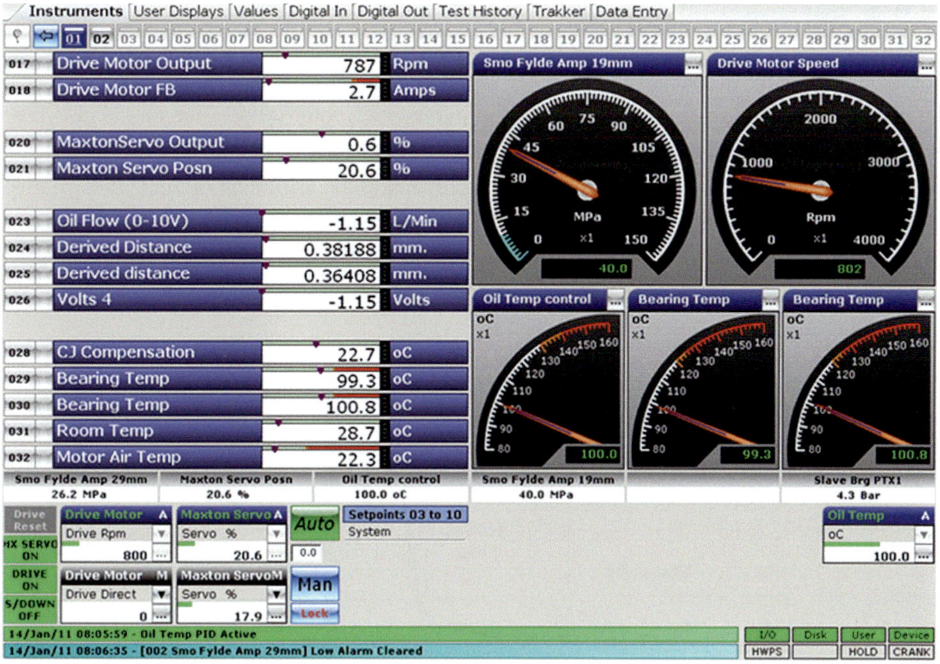

**Figure 3.10:** Control of the automated "Sapphire" test bench

forms cycles consisting of an increase in speed, followed by a stabilization phase, and then reduces the speed back to 0. This cycle is followed for a defined number of repetitions. The change in speed causes a change in the lubrication regime, thus causing accelerated wear. The test bench is fully instrumented and includes a capacitive wear sensor for measuring wear data (**Figure 3.11**).

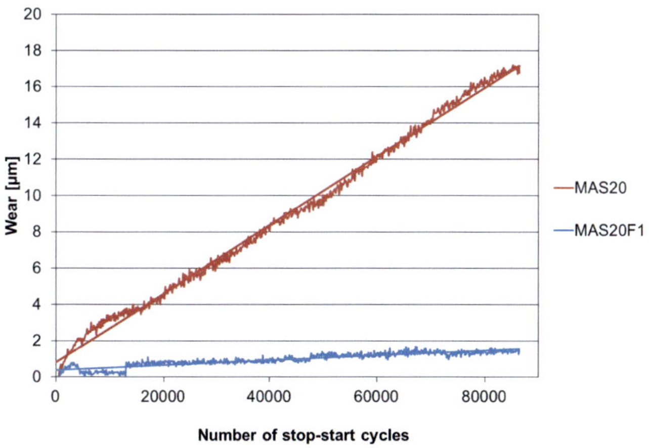


**Figure 3.11:** "Sapphire" test–measuring wear over a number of stop-start cycles performed

### 3.2.5 Seizure resistance

One potential way to evaluate the seizure resistance of various bearing materials is a test in which the formation of a lubricating film is intentionally interrupted. This test is also performed using an automated "Sapphire" test bench by applying a linearly increasing load. The test shaft also has a slot in the axial direction, preventing the formation of the hydrodynamic lubricating film. The test bench is equipped with appropriate instrumentation in order to detect the time of failure of the bearing. The applied load, the first seizure event, and the start of seizure are recorded (**Figure 3.12**).

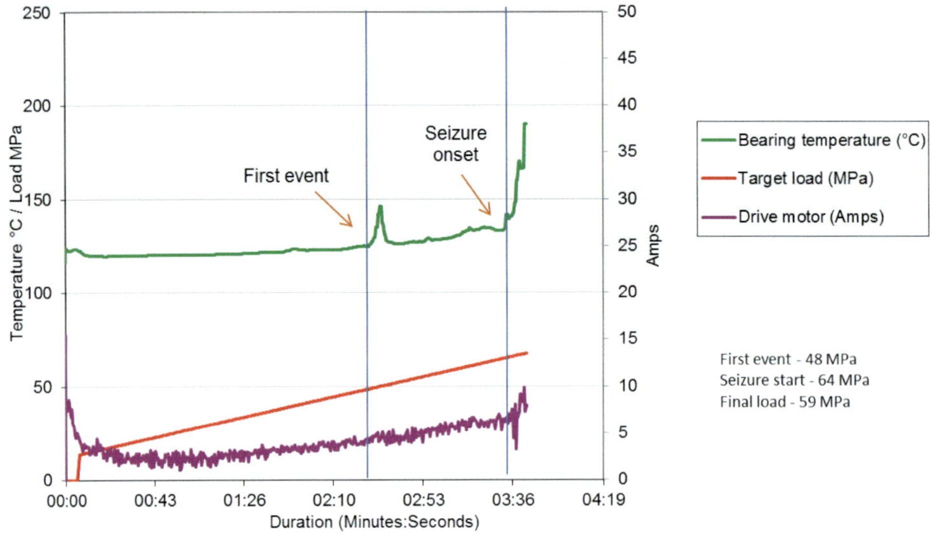


**Figure 3.12:** "Sapphire" seizure resistance test

### 3.2.6 Embeddability

The embeddability of a bearing material is tested by feeding particles of a defined size and hardness to the bearing. A lubricant is contaminated with a defined quantity of particles, which are embedded into the bearing surface by the weight of a ballast (**Figure 3.13**). An imprint (**Figure 3.14**) is then prepared in which the ferrous particles are made visible chemically. The imprint is scanned and digitized. An image processing algorithm is used to evaluate the size, number, and total surface area of the embedded particles. The results are then used to compare various bearing materials using an embeddability index.

**Figure 3.13:** Embeddability test bench

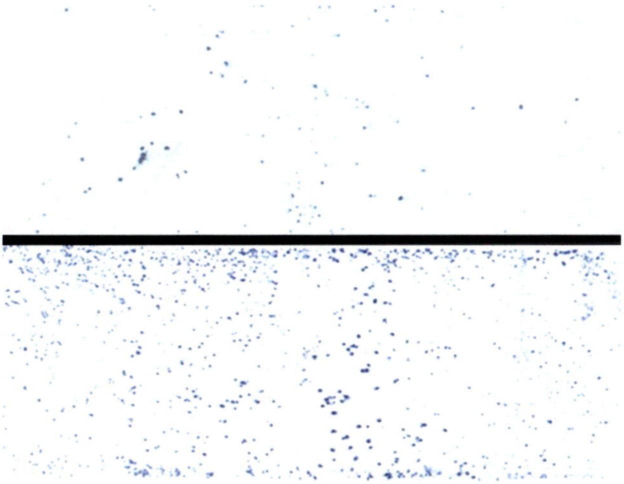

**Figure 3.14:** Examples of iron imprints of the bearing surface after particle embedding

# 3.3 Bearing geometry

## 3.3.1 Bearing diameter and length

The variables of peak oil film pressure (POFP) and minimum oil film thickness (MOFT) are strongly associated with the bearing diameter and bearing length. The length/diameter ratio L/D influences the operating characteristics of the bearing. A larger bearing length reduces the peak pressure in the oil film and increases the minimum oil film thickness. A larger diameter has the same effect. For a given projected bearing surface, the bearing with the higher L/D ratio experiences lower oil film pressures, greater minimum oil film thicknesses, and thus more advantageous load conditions.

### 3.3.2 Grooves and bores

The lubricating oil enters the bearing through grooves and boreholes. Independently of this, they also have a significant influence on the function of the bearings. They are undesirable in loaded areas, because they reduce the usable contact surface of the bearing and thus increase the peak oil film pressure and reduce the minimum oil film thickness. If the grooves and bores are poorly located, there is an increased risk of surface contact between the sliding partners or cavitation damage to the bearing material.

### 3.3.3 Bearing clearance

Bearing clearance has a twofold effect on the properties of the oil film. With less clearance, the loads are better distributed, because the elastic journal deformation that occurs during operation is nearly identical to the bearing curvature and generates a lower peak oil film

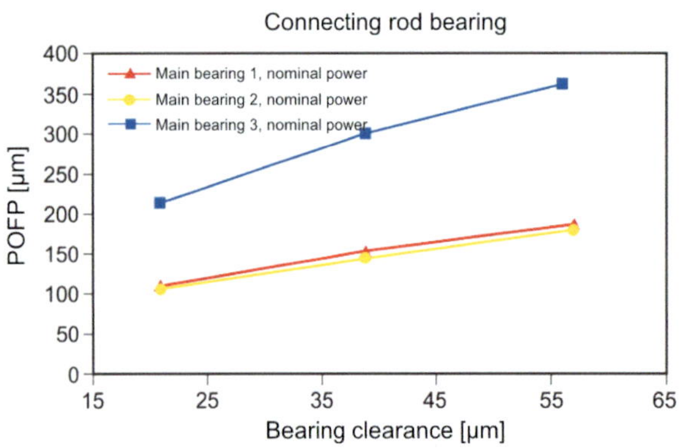


**Figure 3.15:**
Peak oil film pressure POFP as a function of bearing clearance at various rated power levels

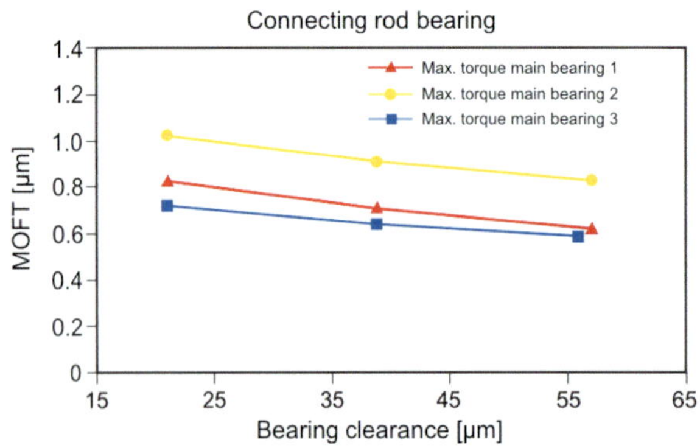


**Figure 3.16:**
Minimum oil film thickness MOFT as a function of bearing clearance

pressure. Lower clearances also generate more heat, which reduces the oil viscosity. The peak oil film pressure POFP increases more or less proportionately with greater clearance (**Figure 3.15**), and the minimum oil film thickness MOFT decreases (**Figure 3.16**).

## 3.3.4 Fit of bearings and bushings

A properly designed fit of the bearing in its housing ensures a reliable seat and good heat transfer due to radial tension. This is achieved through correct design of the bearing overlap. For bearings, this overlap results from the protrusion of the joint face height beyond the housing radius. For bushings, it is the difference in diameter between the bushing and the bore (**Figure 3.17**).

In the past, limit samples with maximum and minimum overlap were prepared, assembled, and measured experimentally in order to validate the design. Today this adaptation is done much more quickly using appropriate computation methods (see Section 3.4.4).

### Eccentricity

Bearing eccentricity is the difference between the vertical and the horizontal diameter. The eccentricity helps to generate sufficient oil film thicknesses, but also helps prevent greater contact load between the journal and the bearing surface when the connecting rod approaches the partition line during the idle phase of the combustion cycle. A simulation of the elasto-hydrodynamic lubrication (EHL), using a special analysis program, allows the selection of the optimal eccentricity for each application.

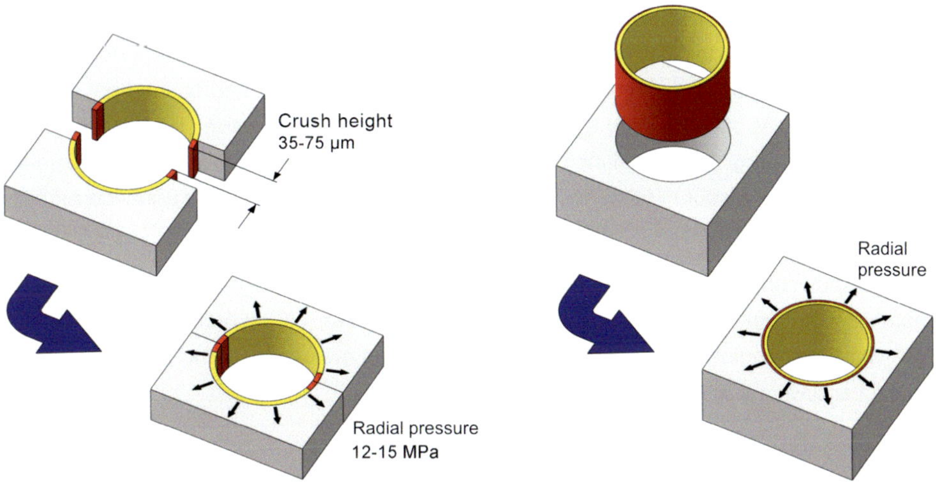


**Figure 3.17:** Definition of fit of half bearing shells and bushings

## 3.4 Numerical simulation

In the development of an engine component, time and costs play an important role. For this reason, a great deal of effort is invested in analysis methods during development, in order to evaluate components and adapt them, on the basis of the results, prior to starting tests. A software package named SABRE (Software for Analysis of Bearings in Reciprocating Engines) has been developed in-house for simulating the behavior of bearings, bushings, and thrust washers in conjunction with assembly and operating parameters. Two main areas of application are differentiated:

- "Routine" simulations for rapid analysis of bearing applications (calculation times from seconds to minutes)
- "Specialized" simulations for detailed analysis of bearing applications (calculation times of hours, days, or weeks)

In order to benefit from the simulation experience (e.g., to establish guidelines), the simulation results are saved in a database (SABRE-DB) and then used to validate new designs by comparison with known solutions.

### 3.4.1 Hydrodynamic lubrication (mobility method)

In addition to the load calculation, the motion of the journal in the bearing is simulated. For this purpose, the two-dimensional Reynolds equation is solved numerically using the finite difference method. The results are then summarized in numerical fit curves using the mobility method. The most important simplification in this case is the assumption of a rigid, cylindrical housing. The main results of this simulation are the maximum specific bearing load (MSL), the minimum oil film thickness (MOFT), and a factor indicating the contact intensity under various operating conditions (PeakDCR Severity). The data required for performing the analysis are the operating parameters of the engine, the crankshaft and bearing geometries, and the properties of the lubricant, which depend heavily on the effective operating temperatures of the bearing. A heat balance (**Figure 3.18**) is therefore required for any bearings for which solutions are sought using the iterative application of the Reynolds equation mobility method.

The computation results can be presented in the form of polar diagrams for the loads and journal orbit diagrams. In addition, diagrams for analysis of the oil film pressure and thickness over the entire engine cycle can be produced, allowing evaluation of the potential risks of contact and wear (**Figure 3.19**).

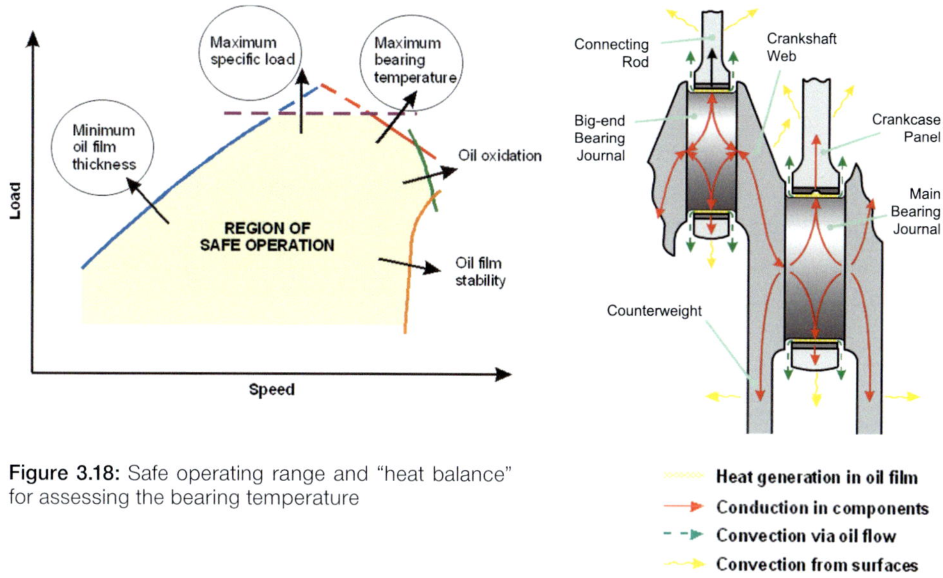


**Figure 3.18:** Safe operating range and "heat balance" for assessing the bearing temperature

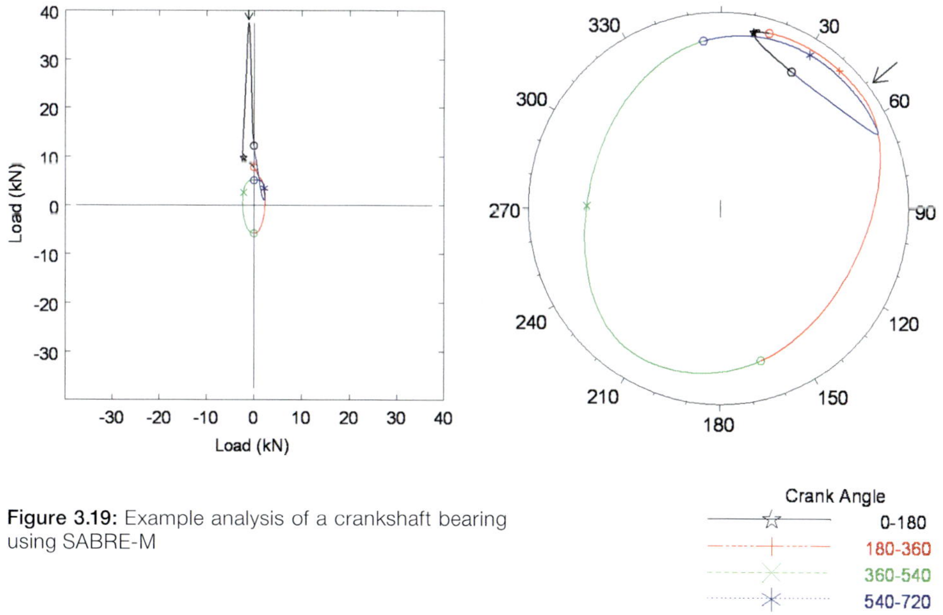


**Figure 3.19:** Example analysis of a crankshaft bearing using SABRE-M

### 3.4.2 Specialized simulations (TEHL)

To obtain more precise results, the same model is used for simulating hydrodynamic lubrication, but with the deformation of the housing due to the bearing load and the heating due to shear work in the oil film (thermo-elasto-hydrodynamic lubrication or TEHL) taken into consideration. The stiffness of the crankshaft, the housing, and the housing shape are determined using a finite element model and also entered into the program. This allows even more detailed results to be obtained for the oil film thickness and the peak oil film pressure (**Figure 3.20**).

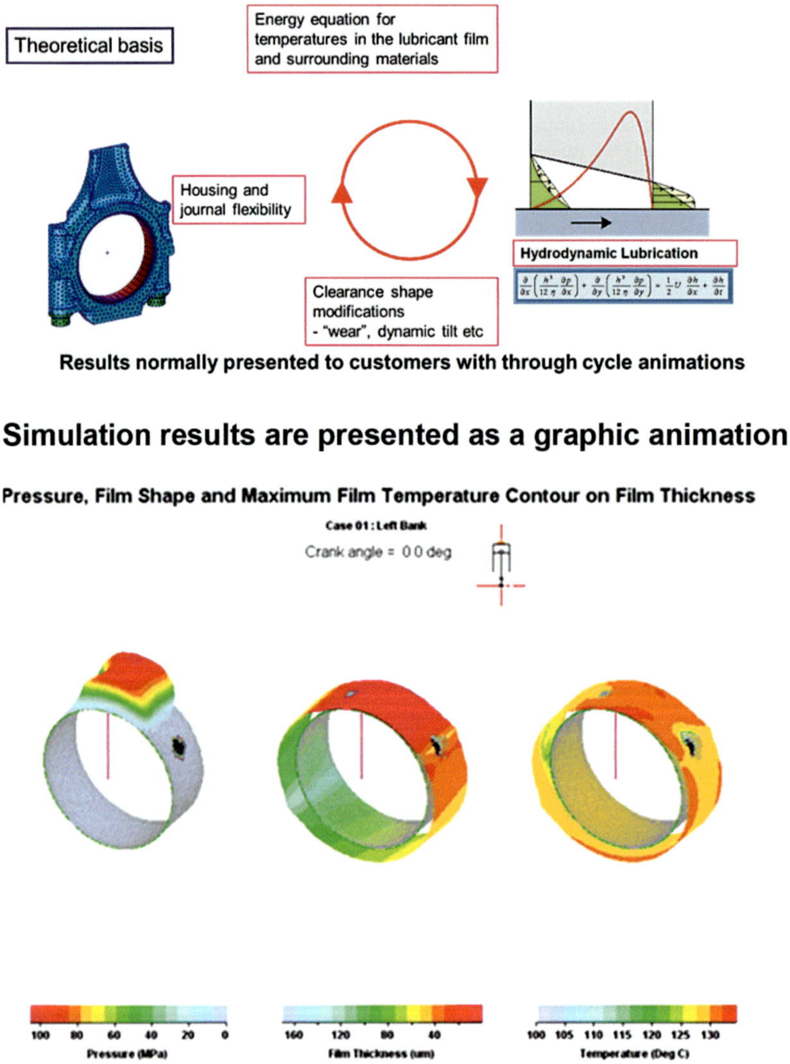


**Figure 3.20:** Summary of a SABRE-TEHL computation and example of animation of the oil film pressure, shape, and temperature

The use of the elasto-hydrodynamic theory assumes lubrication, which takes into consideration not only the hydrodynamic pressure but also the metal-to-metal contact pressure.

Evaluation criteria for these computation results include wear, peak oil film pressure, power loss, oil flow, and maximum temperature.

### 3.4.3 Additional CFD simulations

In addition to the previously described TEHL method, more advanced computations are sometimes needed in order to better understand the environment outside of the bearing clearance and the influences of materials. One important tool in this context is computational fluid dynamics (CFD) for computing the oil flow out of the gallery, through the oil grooves of the main bearing, and through the crankshaft bores to the connecting rod bearing in the big end bore. At high engine speeds or low supply pressure in the gallery, for example, bubbles can form in the crankshaft bores and cause the big end bore of the conrod to be undersupplied with lubricant. CFD technology is also helpful for evaluating various groove geometries and assessing the risk of seizure due to undersupply of lubricant (**Figure 3.21**).

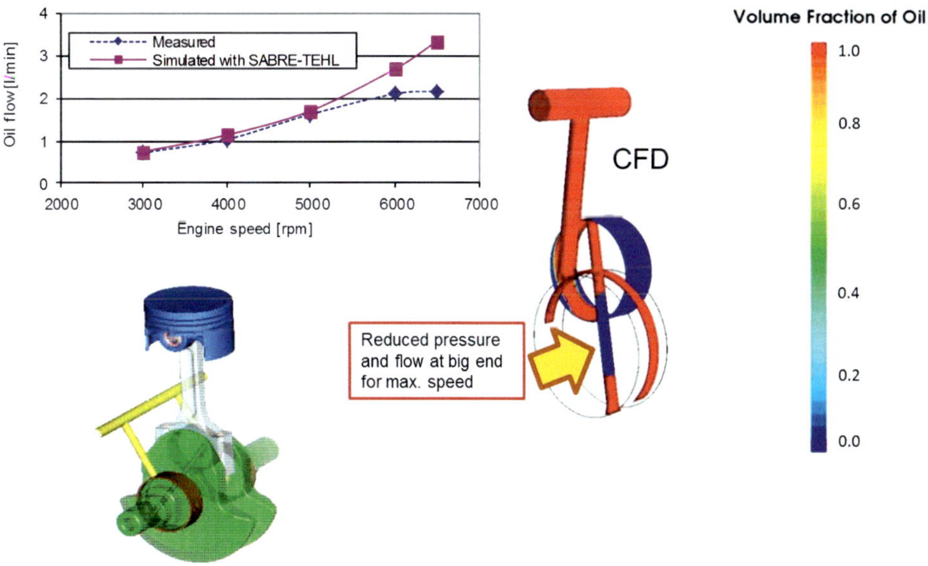


**Figure 3.21:** CFD computation for testing oil transport between bearing locations–bubble formation in the crankshaft bore

### 3.4.4 Interference and assembly simulations

The behavior of the bearings and bushings depends on how securely these components are installed in their housings. A proper fit ensures that the component is held securely and provides appropriate heat transfer and optimal bearing clearance. The routine simulation is based on the theory of solid cylinders and uses automated finite element analysis (FEA) within a customized analysis program named SABRE-FIT-FEA. The data entered consist of the geometric features of the assembly and the housing, the properties of the bearing material, and the operating temperatures. The results are stresses and diametric overlaps or clearances at different temperatures.

A special FEA simulation can also be used for main bearings in order to investigate the influence of housing oil grooves and the engine block (**Figure 3.22**).

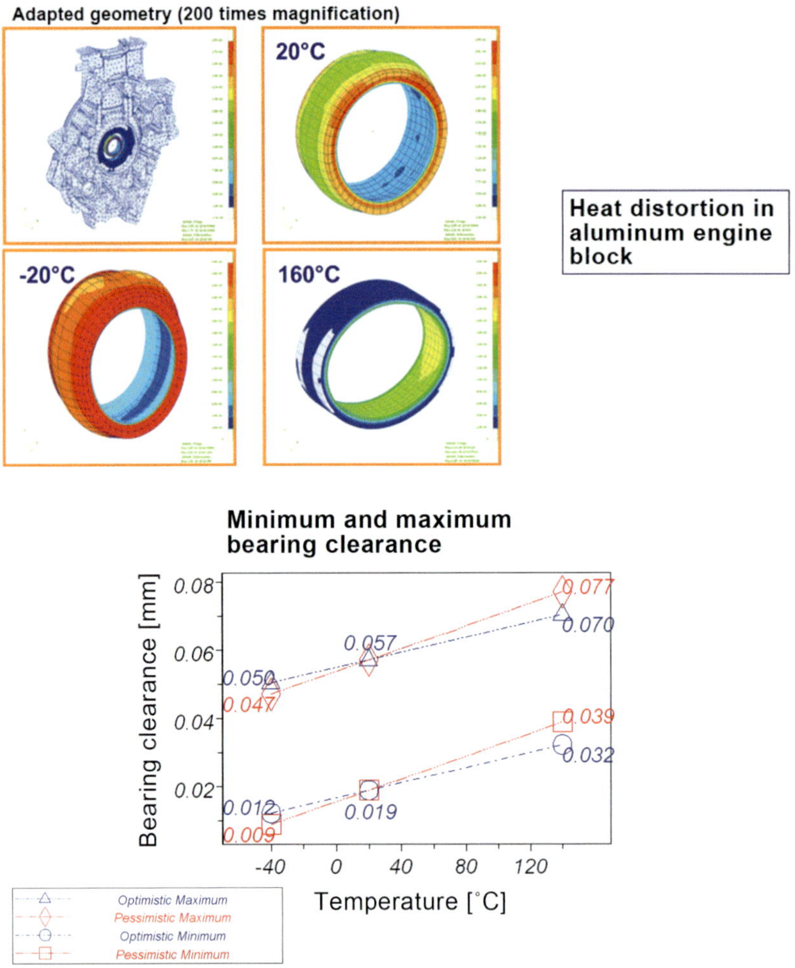


**Figure 3.22:** Illustration of a routine and specialized simulation of the assembly of main bearings

# 3.5 Materials

Selection criteria for bearing materials include the load and the permissible stress of the material. The load carrying capacity limits are determined for each material on the basis of simulations, bench tests, and engine testing. They are lower for main bearings, because of potential alignment errors.

For axial bearings, the selection of the material is based on empirical analysis, considering the geometric and material factors.

## Composition and properties of bearing materials

**Table 3.4:** Aluminum alloys

| Description | Chemical composition of the core alloy [%] | | | | | Application/ properties | Process | Minimum hardness (alloy/steel support shell) | Specific bearing load carrying capacity [MPa] | Design |
|---|---|---|---|---|---|---|---|---|---|---|
| | Al | Sn | Si | Cu | Other | | | | | |
| MAS 19 | 89 | 6 | 2 | 1 | Ni 1<br>Mn < 1<br>V < 1 | Plain bearings, bushings, and thrust washers; medium load carrying capacity with high wear resistance and embeddability | Cast aluminum alloy roll-bonded on steel | MAS 20: 45–62 $HV_{1-5}$<br>Steel: 155–235 $HV_{10}$ | 65 | Bimetal material with fine formation of the tin phase in an aluminum matrix, combined with an aluminum interlayer and roll-bonded onto a low-carbon steel support shell |
| MAS 20 | 89 | 6 | 2 | 1 | Ni 1<br>Mn < 1<br>V < 1 | Plain bearings, bushings, and thrust washers; high load carrying capacity with high wear resistance and embeddability | Cast aluminum alloy roll-bonded on steel | MAS 20: 58–72 $HV_{1-5}$<br>Steel: 155–235 $HV_{10}$ | 75 | Bimetal material with fine formation of the tin phase in an aluminum matrix, combined with an AlCu interlayer and roll-bonded onto a low-carbon steel back |

**Table 3.5:** Alloys of cast bronze (overlays, see **Table 3.8**)

| Description | Chemical composition of the core alloy [%] | | | | Application/ properties | Process | Minimum hardness (alloy/steel support shell) | Specific bearing load carrying capacity [MPa] | Design |
|---|---|---|---|---|---|---|---|---|---|
| | Cu | Pb | Sn | Other | | | | | |
| MCB 1 | 78 | 20 | 2 | | Bearing material for trimetal bearings | Lead-bronze alloy cast on steel | MCB 1: 70 – 110 $HV_5$ Stahl: 121 – 195 $HV_{10}$ | See upper limit of overlay | Bimetal material, copper-tin base material, cast on steel |
| MCB 2 | 75 | 23 | 2 | | Bearing material for trimetal bearings | Lead-bronze alloy cast on steel | MCB 2: 40 – 95 $HV_5$ Stahl: 115 – 230 $HV_{10}$ | See upper limit of overlay | Bimetal material, copper-tin base material, cast on steel |
| MCB 5 | 80 | 10 | 10 | | Bearing material for rod bushings | Lead-bronze alloy cast on steel | MCB 5: 95 – 160 $HV_{2,5-5}$ Stahl: 90 – 215 $HV_{10}$ | 130 | Bimetal material, copper-tin base material, cast on steel |
| MCB 17 | 91 | | 4 | Bi 4 Ni 1 | Lead-free bearing material for trimetal bearings with galvanically applied overlay or polymer overlay | Lead-free bronze cast on steel | MCB 17: 76 – 125 $HV_5$ Stahl: 145 – 240 $HV_{10}$ | See upper limit of overlay | Bimetal material with bismuth, copper-tin base material, cast on steel |
| MCB 18 | 95 | | 4 | Bi 1 | Lead-free bearing material for trimetal bearings with sputter overlay | Lead-free bronze cast on steel | MCB 18: 80 – 140 $HV_5$ Stahl: 120 – 220 $HV_{10}$ | See upper limit of overlay | Bimetal material with bismuth, copper-tin base material, cast on steel |
| MCB 20 | 91 | | 8 | Ni 1 | Lead-free bearing material for rod bushings | Lead-free bronze cast on steel | MCB 20: 120 – 195 $HV_{2,5-5}$ Stahl: 100 – 180 $HV_{10}$ | See upper limit of overlay | Bimetal material, copper-tin base material, cast on steel |
| MCB 25 | 87 | | 8 | Bi 4 Ni 1 | Lead-free bearing material for rod bushings with improved corrosion and seizure resistance | Lead-free bronze cast on steel | MCB 25: 120 – 195 $HV_{2,5-5}$ Stahl: 100 – 180 $HV_{10}$ | See upper limit of overlay | Bimetal material, copper-tin base material, cast on steel |

**Table 3.6:** Sintered bronze alloys

| Description | Chemical composition of the core alloy [%] | | | | Application/ properties | Process | Minimum hardness (alloy/steel support shell) | Specific bearing load carrying capacity [MPa] | Design |
|---|---|---|---|---|---|---|---|---|---|
| | Cu | Pb | Sn | Other | | | | | |
| MSB 10 | 80 | 10 | 10 | | Standard bronze for bushings | Lead-bronze alloy sintered on steel | MSB 10: 70–165 $HV_5$ Steel: 105–165 $HV_{10}$ | 130 | Bimetal material with consistently formed lead phase, copper-tin base material, sintered on steel |
| MSB 20 | 91 | | 8 | Ni 1 | Sintered lead-free material for bushings with increased corrosion resistance | Lead-free bronze alloy sintered on hard steel | MSB 20: 73–87 $HV_5$ Steel: 105–165 $HV_{10}$ | 150 | Lead-free copper-tin Bimetal material, sintered on steel |
| MSB 21 | 91 | | 8 | Ni 1 | Sintered lead-free trimetal bearing material for polymer bearing layers | Lead-free bronze alloy sintered on hard steel | MSB 21 95–200 $HV_5$ Steel: 170–220 $HV_{10}$ | See upper limit of overlay | Lead-free copper-tin Bimetal material, sintered on hard steel |
| MSB 30 | 87 | | 8 | Bi 3 Ni 1 $Al_2O_3 < 1$ | Sintered lead-free material for bushings with increased corrosion and wear resistance | Lead-free bronze alloy sintered on hard steel | MSB 30: 95–200 $HV_5$ Steel: 105–165 $HV_{10}$ | 150 | Lead-free copper-tin Bimetal material, sintered on steel |

**Table 3.7:** Galvanic bearing layers (overlays)

| Description | Chemical composition of the core alloy [%] | | | | | | Application/ properties | Process | Specific bearing load carrying capacity [MPa] | Design |
|---|---|---|---|---|---|---|---|---|---|---|
| | Pb | Sn | Cu | In | Al | Other | | | | |
| P 3 | 87 | 10 | 3 | 1 | | | Lead-based overlay for less demanding applications | Galvanic application | 70 | Galvanically applied lead layer with homogeneously distributed copper-tin; with nickel interlayer |
| P 5 | 85 | 10 | 5 | | | | Lead-based overlay with improved wear resistance | Galvanic application | 75 | Galvanically applied lead layer with homogeneously distributed copper-tin; with nickel interlayer |

| Description | Chemical composition of the core alloy [%] | | | | | | Application/ properties | Process | Specific bearing load carrying capacity [MPa] | Design |
|---|---|---|---|---|---|---|---|---|---|---|
| | Pb | Sn | Cu | In | Al | Other | | | | |
| P 9 | 81 | 10 | 9 | | | | Higher load carrying capacity for lead-based overlays | Galvanic application | 80 | Galvanically applied lead layer with homogeneously distributed copper-tin; with nickel interlayer |
| Q 1 | 92 | | | 8 | | | High-performance gasoline and diesel engines | Galvanic application | 85 | Galvanically applied lead layer with homogeneously distributed indium |
| C 1 | 88 | 11 | | | | $Al_2O_3$ 1% | Overlay with increased wear resistance for passenger car diesel engines | Galvanic application | 80 | Galvanically applied lead layer with homogeneously distributed aluminum oxide and local tin enrichment; with nickel interlayer |
| C 2 | 75 | 10 | | 14 | | $Al_2O_3$ 1% | Higher load carrying capacity and wear resistance for passenger car diesel engines | Galvanic application | 90 | Galvanically applied lead-indium layer with homogeneously distributed aluminum oxide and local tin enrichment; with nickel interlayer |
| T 5 | | 99 | | | | 1 | Lead-free galvanically applied layer for high-load applications | Galvanic application | 85 | Galvanically applied tin layer with fine-grained structure; with nickel interlayer |

**Table 3.8:** Polymer bearing layers (overlays)

| Description | Chemical composition of the core alloy [%] | Application/properties | Process | Specific bearing load carrying capacity [MPa] | Design |
|---|---|---|---|---|---|
| F1,<br>F2,<br>F3 | PAI polymer matrix<br>Metal flakes<br>Solid lubricant | Lead-free polymer bearing layer for stop-start applications with high wear resistance and seizure resistance | Polymer | 85–105 | Polymer layer with a homogeneous distribution of metal flakes and solid lubricant |

**Table 3.9:** Sputter bearing layers (overlays)

| Description | Chemical composition of the core alloy [%] | | | | | | Application/ properties | Process | Specific bearing load carrying capacity [MPa] | Design |
|---|---|---|---|---|---|---|---|---|---|---|
| | Pb | Sn | Cu | In | Al | Other | | | | |
| S1 | | 40 | 1 | | 59 | | Sputter overlay for passenger car diesel engines | Sputter | 110 | Sputter aluminum-copper layer with fine homogeneously distributed tin phase; with nickel-chromium interlayer |
| S2 | | 30 | 1 | | 69 | | Sputter overlay for high-performance passenger car applications | Sputter | 130 | Sputter aluminum-copper layer with fine homogeneously distributed tin phase, with nickel-chromium interlayer |
| S3 | | 40 | 1 | | 59 | | High-performance passenger car diesel engines | Sputter | 130 | Sputter aluminum-copper layer with fine homogeneously distributed tin phase, with aluminum-tin interlayer |
| S10 (1st intermediate layer–2nd intermediate layer-bearing layer) | | 6 | 1 | | 90 | Ni 1 Si 2 | Top-performance diesel engines for passenger cars | Sputter | 130 | Graded sputter aluminum-copper layer with fine homogeneously distributed tin phase |
| | | 30 | 1 | | 67 | Fe 2 | | | | |
| | | 30 | 1 | | 69 | | | | | |

# 3.6 Market requirements and technology trends

The goals of ongoing development of engines are higher specific power output, lower fuel consumption, lower emissions, smaller designs, and lower costs. These result in greater demands on MAHLE engine components in terms of wear resistance, load carrying capacity, and seizure resistance. **Table 3.10** shows a summary of the effects of these goals on the bearing portfolio.

**Table 3.10:** Market demands and regulatory goals for engine components

| | Engine trends | Effects on the operating characteristics of the engine | Effects on bearings |
|---|---|---|---|
| **1. Regulatory goals** | | | |
| Emissions and particle reduction | Reduction of engine friction | Lower oil viscosity Stop-start | Increased wear, redesign |
| | Reduction of engine weight | Lighter components, aluminum crankcase | Excessive housing deformation |
| | Gasoline direct injection | Higher piston weight | Greater inertial load |
| | Exhaust gas recirculation | Oil contamination | Increased wear |
| | Increase in peak cylinder pressure | Greater mechanical loads | Greater loads |
| Noise | Less vibrations | Reinforced crankcase | Housing adaptation |
| Prohibited materials | Lead-free components | | Lead-free materials |
| **2. Customer demands** | | | |
| Higher performance | Greater air consumption and greater blow-by | Higher temperatures and engine speed | Overheating, higher inertial loads |
| | Increase in peak cylinder pressure | Greater mechanical loads | Greater loads |
| Lower fuel consumption | Reduced engine friction, downsizing | Lower oil viscosity | Increased wear, redesign |
| | Reduction of engine weight | Lighter components, aluminum crankcase | Excessive housing deformation |
| | Gasoline direct injection | Higher piston weight | Increased inertial loads |
| Longer oil change intervals | | Oil contamination and aging | Increased wear, corrosion |
| Service life, reliability | Higher vehicle miles traveled | | Redesign |

To meet these demands, the bearings were adapted in terms of dimensions and materials. High-strength aluminum alloys for bimetal bearings are newly developed for this purpose. New bearing layers for trimetal bearings and lead-free materials as a substitute for the traditional leaded bronze have also been introduced on the market. In the future, great additional potential can be expected from the new polymer bearing layers in particular.

# 4 Connecting rod

## 4.1 Introduction

The connecting rod connects the piston to the crankshaft and consists of the small end and big end bores as well as the shank.

The rotation of the crankshaft induces a rotational motion of the big end bore, which has a bearing eccentric to the axis of the crankshaft. The small end bore follows the axial stroke motion of the piston in the cylinder (**Figure 4.1**). The connecting rod is thus a machine element that transforms the axial motion of the piston into the rotation of the crankshaft.

The space covered by the connecting rod during one revolution of the crankshaft, also known as the conrod sweep (**Figure 4.2**), must be considered in collision studies for the crankcase and engine block.

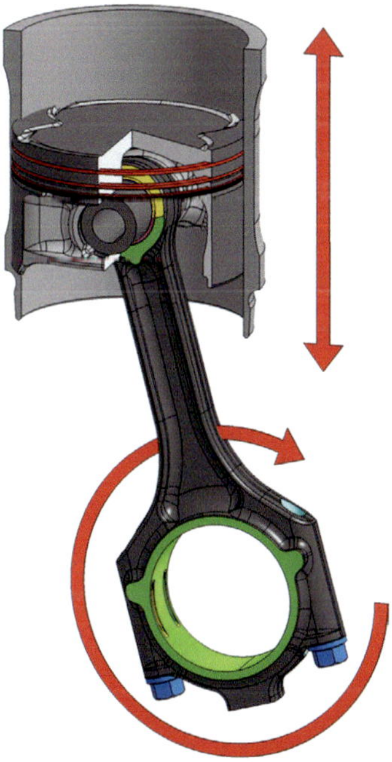

**Figure 4.1:** Main motions of the piston-connecting rod system (vertical arrow: oscillating; circular motion: rotational)

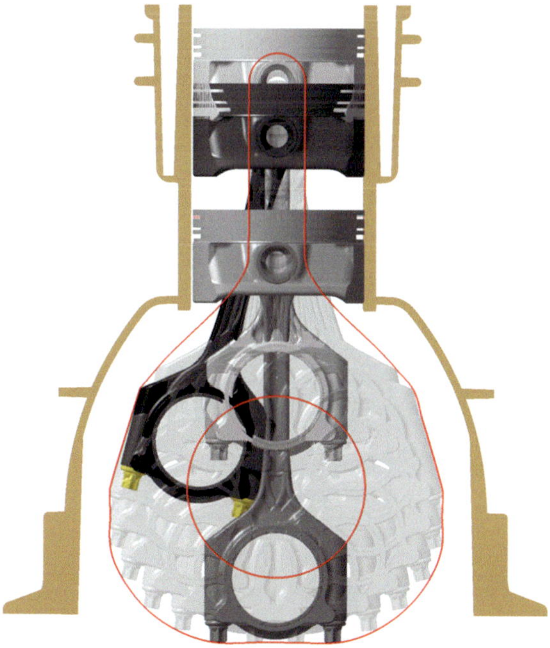

**Figure 4.2:** Conrod sweep

While the small end bore is always closed, the big end bore is normally designed to come apart for assembly. **Table 4.1** provides information about the different design details of connecting rods, but not about the interrelationship of individual details. The task of the designer is to determine the correct configuration associated with the requirements profile.

**Figure 4.3** shows the important terminology and dimensions of a connecting rod.

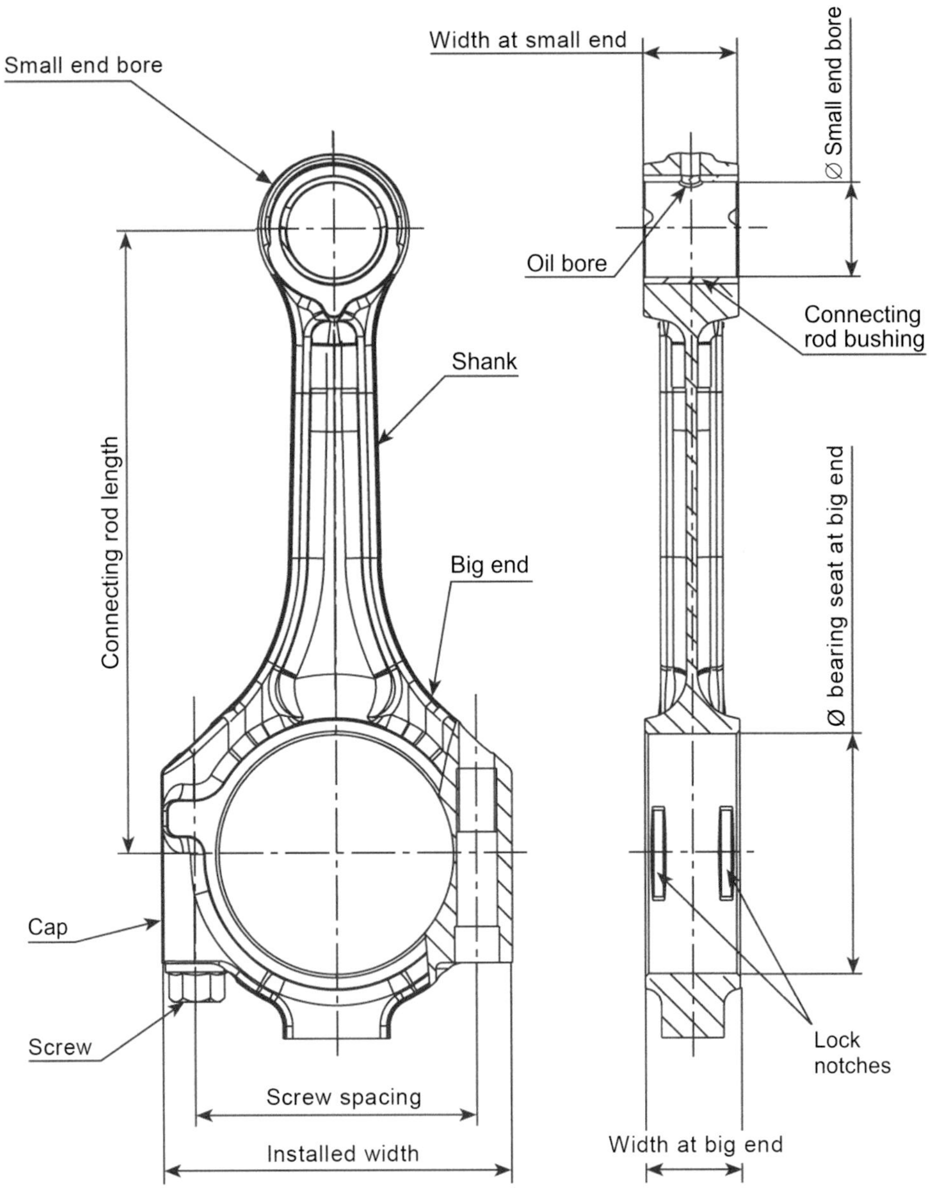


**Figure 4.3:** Terminology and major dimensions of a connecting rod

**Table 4.1:** Types of connecting rods and design parameters

<table>
<tr><th>Area</th><th colspan="5">Type</th></tr>
<tr><td>Small end bore</td><td>Parallel</td><td colspan="3">Stepped</td><td>Keystone</td></tr>
<tr><td>Piston pin (small end bore)</td><td colspan="2">Floating</td><td colspan="3">Fixed</td></tr>
<tr><td>Shank</td><td colspan="2">I-section</td><td colspan="3">H-section (motorsport)</td></tr>
<tr><td>Big end bore</td><td colspan="2">Straight-split</td><td colspan="3">Angle-split</td></tr>
<tr><td>Parting plane of big end bore</td><td>Cracked</td><td colspan="3">Machined flat, with dowel sleeve/fitting screw/dowels</td><td>Tooth profile</td></tr>
<tr><td>Blank production</td><td>Forging</td><td colspan="2">Casting</td><td colspan="2">Powdered metal/sintering</td></tr>
</table>

## 4.2 Stresses

As the element that transfers forces and motions between the piston and the crankshaft, the connecting rod is subjected to large, alternating loads. The connecting rod is loaded by the piston in compression (under prevailing gas force) and in tension (primarily because of inertia force). The connecting rod is also stressed in bending as a result of its pivoting motion. As a moving engine component, it should be as light as possible and sufficiently stiff in shape in terms of interacting with the piston pin and the crankshaft pin. Sufficient component and structural strength must also be ensured.

The transmission of power from the piston and piston pin via the connecting rod to the crankshaft is achieved by the lubrication in the bearings. The force applied to the connecting rod is therefore dependent on the pressure distribution in the lubricant. This, in turn, is affected by the stiffness of the conrod bores. The inertia force is held in equilibrium by the lubrication pressure between the crankshaft pin and the cap side bearing. The force flow between the connecting rod and the bearing cap is provided by the connecting rod bolts. The conrod bore deforms under inertia force with ovality in the vertical direction and the bolts are bent outward. If the bolt force is insufficient, the connecting rod joint will gape on the inside; see **Figure 4.4**.

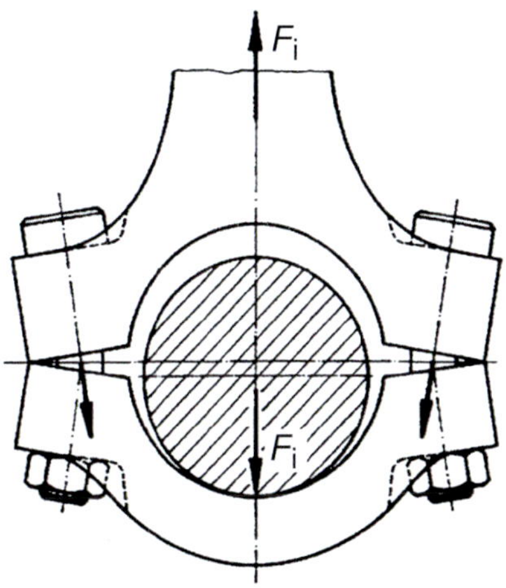


**Figure 4.4:**
Deformation of the big end bore

Under maximum gas pressure, however, the connecting rod shank presses on the crankshaft pin via the lubrication. The connecting rod bore becomes transversely oval and the bolts bend inward. As a result of these deformations, considerable bending stress occurs in the conrod bores. The most highly stressed areas in straight-split connecting rods, in addition to the bolt threads, are the fillets on the transition from the shank to the big end bore and to the small end bore. With angle-split connecting rods, the upper part of the blind hole thread is located directly in the force flow, which leads to a stress peak (**Figure 4.7**).

## 4.3 Requirements

### Mass of the connecting rod

As a general principle, moving masses should be kept as small as possible, in order to help keep fuel consumption low and to reduce vibration excitation. Weight can be saved in the engine because of the lower overall height resulting from reduced connecting rod length. The changes to the lateral forces on the piston skirt, however, must be taken into consideration.

In order to maintain high running smoothness and low vibration levels, the rotating and oscillating masses should match as closely as possible among the individual cylinders. The oscillating mass portion is located on the piston side and the rotating portion is on the crankshaft side. There exist several potential ways to attain this goal.

The sintering method allows tolerances in raw sintered weight with a spread of less than 1%. MAHLE has comprehensively developed industrial engineering for forging connecting rods and significantly reduced weight variation. The controlled, fully automated forging process thus allows a spread of less than 1% in the raw forging weight.

Another option is classification. The oscillating and rotating masses of the finished connecting rods are determined and the connecting rods are divided into different weight classes. For this purpose, the connecting rod is weighed horizontally with two scales, each at the center point of the small end and big end bores. The value at the small end bore corresponds to the oscillating mass, and that of the big end bore to the rotating mass (**Figure 4.5**).

When machining to weight, a weight slug is added on the big end bore (sometimes on the small end as well), which is milled off to adjust to the desired weight.

Only one connecting rod weight class is installed in a given engine. Because different diameter classes are often required for the piston, depending on the finished cylinder diameter, the assembly unit consisting of the piston, piston rings, piston pin, circlips, and connecting rod is assembled directly in the engine for installation.

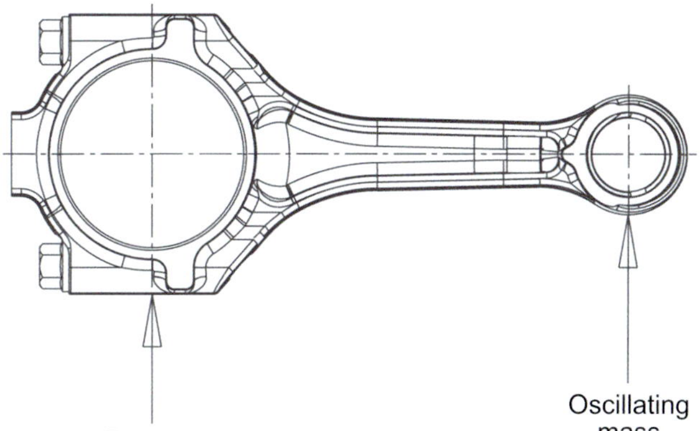


**Figure 4.5:**
Distribution of moving masses of a connecting rod

# 4.4 Big end bore

The diameter of the big end bore is determined from the crankshaft pin diameter of the crankshaft and the bearing wall thickness. The critical stress for the big end bore results from inertia force. The oscillating mass force loads the big end bore in tension, and the bore is ovally deformed along the longitudinal axis of the connecting rod. This results in bending stresses and transverse forces in the parting surface. It is important that the parting line remains closed under all operating points.

## 4.4.1 Cracking (fracture splitting)

Cracking, or fracture splitting, of connecting rods has become common practice in recent years. Nearly all new designs in series production today employ this method to create the parting in the big end bore. The big end bore is notched inside the bore with a laser beam or reamer. For sintered parts, the notch is pressed in during the manufacture of the blank. Using a cracking mandrel, the halves are then broken apart (cracked) hydraulically at room temperature (**Figure 4.6**). The resulting joint face (fracture surface) is not machined, and dowel sleeves or fitting screws are not needed. The fit is provided solely by the engagement of the uneven surfaces. The fracture surfaces experience only minimal settling. In cracking, the fracture splitting ability of the connecting rod material is critical. Special steel materials with a yield point/tensile strength ratio of up to 0.75 are used.

**Figure 4.6:**
Fracture surfaces of the big end bore, manufactured by cracking

## 4.4.2 Angle split of the big end bore

If the crankshaft has a large crankshaft pin diameter, the big end bore must be split at an angle in order to allow the connecting rod to be installed and removed through the cylinder liner. This leads to complex loading conditions in the parting joint.

In an angle-split connecting rod (**Figure 4.7**), the upper blind hole thread is particularly at risk, because it is located directly in the force flow of the entire connecting rod. This is the area of alternating tensile and compressive loads, which are increased further by the notch effect of the thread, resulting in an increased risk of fracture. The cross section around this thread must therefore be dimensioned carefully.

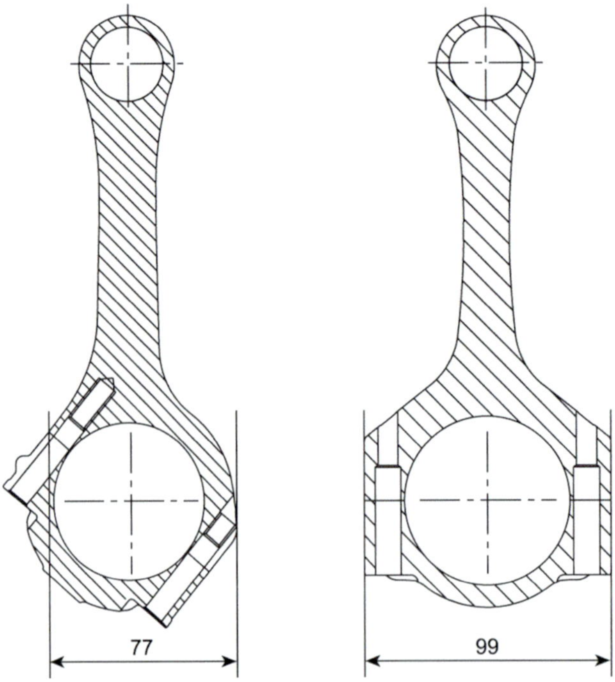


**Figure 4.7:**
Angle- or straight-split connecting rod and required clearance of the cylinder liner for identical crankshaft pin diameter

# 4.5 Connecting rod shank

Looking at the shank cross section in pivoting direction (perpendicular to the crankshaft axis), a differentiation is made between the I- and H-section. The latter is often used in racing engines because of the bending loads at high speeds. The I-section is preferred for mass-production engines because of simpler blank production and thus lower costs at higher quantities.

The connecting rod shank is subjected to an alternating tension/compression load in four-stroke engines (tension due to inertia force at TDC nonfired; compression due to gas force at TDC fired). In addition to fatigue resistance, the connecting rod shank must also feature sufficient buckling resistance.

To supply oil to the small end bore, oil can be fed under pressure from the crankshaft through a bore along the length of the connecting rod shank.

# 4.6 Small end bore

## 4.6.1 Pin bearing in the small end bore

The small end bore accepts the piston pin and, together with the piston pin boss, forms the joint about which the connecting rod pivots.

In a fixed-pin connecting rod, the piston pin is shrink-fit in the small end bore and has clearance only in the piston pin boss. To assemble the piston pin, the small end bore is heated to approximately 400°C. This assembled unit can no longer be disassembled non-destructively.

For a floating piston pin design, press-fit bushings are used in the small end bore. The piston pin has clearance both in the connecting rod and in the piston pin boss. It must be held in the piston axially by piston pin circlips (Chapter 2.8) and can "float" circumferentially on the oil film. The piston boss can withstand higher loads because of the intensive lubrication, or a shorter piston pin can be used with the same load. In highly stressed engines, therefore, the piston pins have floating bearings.

The advantages and disadvantages of fixed-pin connecting rods and floating piston pins are summarized in **Table 4.2**.

**Table 4.2:** Advantages and disadvantages of fixed-pin connecting rods and floating configurations for piston pins

| Fixed pin connecting rod | Floating design |
|---|---|
| **Advantages** | **Advantages** |
| No circlip needed | Assembled unit can be disassembled |
| No sliding bearing needed in the connecting rod, such as a bushing | Lower weight due to greater load carrying capacity |
| **Disadvantages** | **Disadvantages** |
| Piston pin cannot be removed easily | Circlip grooves and circlips must be provided |
| Difficult to assemble piston pin | Locking devices must be assembled |
| Higher weight with longer piston pin, due to lower load carrying capacity | Connecting rod bushings generally required |

## 4.6.2 Geometry of the connecting rod small end

Contact pressure, as a dimension for the bearing load in the small end bore, is derived from the gas force, pin diameter, and bearing length. It is generally greater than 100 MPa.

Gas force, as the highest magnitude load, acts only in the direction of the big end bore, which has led to the development of various types of support in the small end bore to meet requirements relating to load carrying capacity, weight, and cost.

The parallel connecting rod is the basic version and the easiest to use. It is the most economical for manufacturing if the big end bore has the same width as the small end bore. Because of increasing gas pressures, this type is often replaced with one of the variants described below, whereby the lower part of the small end bore, which is subjected to the gas force, is wider than the upper part, which is subjected only to inertia forces (**Figure 4.8**).

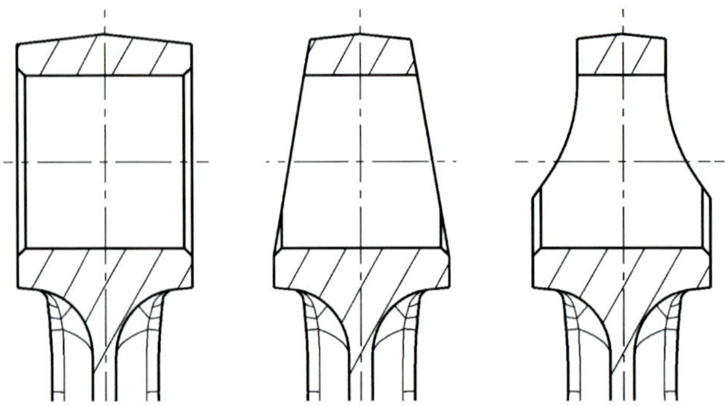

**Figure 4.8:**
Cross-sectional shapes of the small end bore

Left: parallel connecting rod

Middle: tapered connecting rod

Right: stepped connecting rod

The tapered connecting rod is slanted at the small end bore and thus gets wider in the direction of the connecting rod shank. The piston pin boss is adapted accordingly, in order to reduce contact pressure here as well. During design, particular care must be taken to ensure that the distance between the connecting rod and the piston pin boss in the direction of the piston pin axis is reduced when the connecting rod pivots (**Figure 4.8**).

The stepped connecting rod presents the greatest challenge to the manufacturing process. However, it best combines load carrying capacity and shape. During design, here again, it must be ensured that the connecting rod does not collide with the piston when it pivots (**Figure 4.8**).

## 4.6.3 Lubrication of the small end bore

Oil is sprayed onto the bottom side of the piston through spray nozzles in the crankcase in order to dissipate heat absorbed by the piston. This oil drips onto the small end bore and moves laterally or through oil bores into the lubricant gap between the piston pin and the bore.

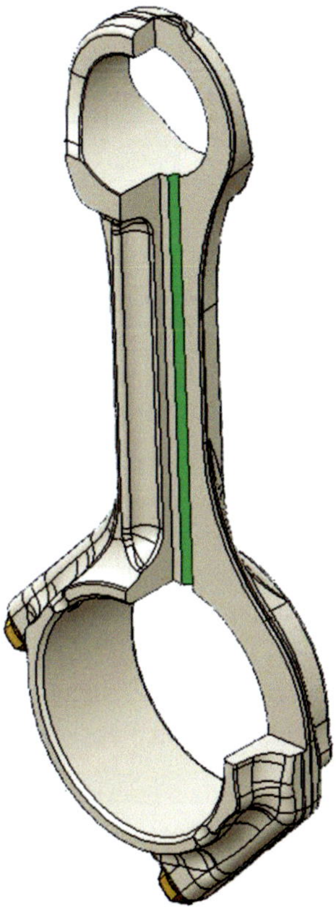

If the amount of lubricating oil coming from the piston is not sufficient, then oil is fed into the lubricant gap, starting from the crankshaft, through a bore in the shank (**Figure 4.9**).

This design is used primarily in connecting rods for commercial vehicle diesel engines, which are run at low speeds and have sufficient space for the oil bore because of the cross-sectional area of the shank. This measure is independent of the use of a bearing bush in the small end bore.

**Figure 4.9:**
Oil bore in the shank

## 4.6.4 Bushingless pin bearing in the small end bore

By eliminating the bearing bush, the wall thickness in the small end bore can be reduced. This solution reduces oscillating masses, which is of increasing importance in terms of running smoothness and fuel consumption. The trend for passenger car engines is therefore toward bushingless pin bearings. Good lubrication of the small end bore is critical for this concept. Various measures are available for improving tribological behavior. They are defined as a function of the contact pressure in the conrod bore:

- The bore is designed as a shaped pin bore, where the bore tapers off to the sides (**Figure 4.10**).This prevents the piston pin, which bends under the maximum gas pressure, from making contact with the edge of the connecting rod and consequently seizing.

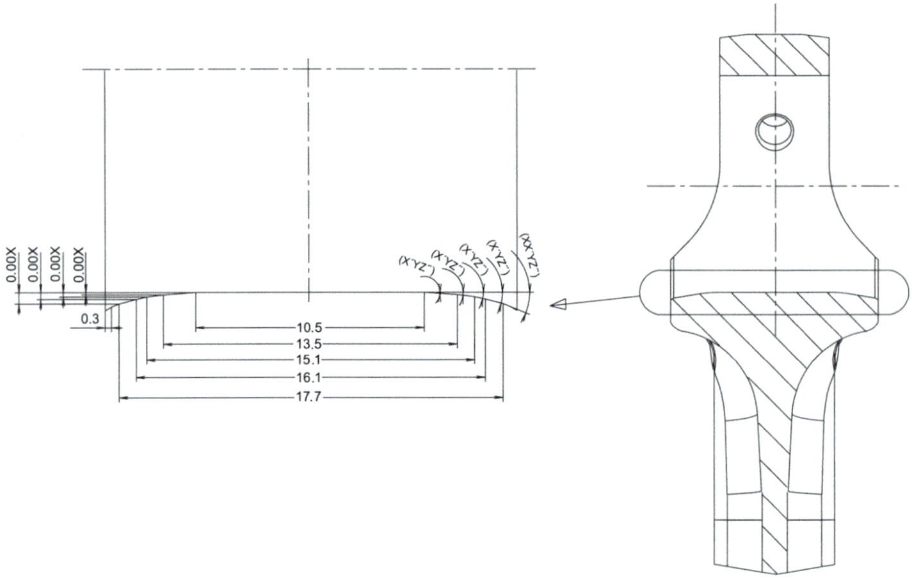


**Figure 4.10:** Shape optimization of the small end bore, without bushing

- Specialized mechanical processing, such as roller-burnishing, can harden the surface and produce a defined surface roughness. The goal is to prevent peaks in the surface structure that would contribute to seizing, and to obtain valleys in which oil collects.
- Another possibility is the use of oil pockets in the area of transition to the bore (**Figure 4.11**). These recesses are already produced within the forging. Oil collects in the pockets and is drawn into the lubricant gap by the capillary effect.

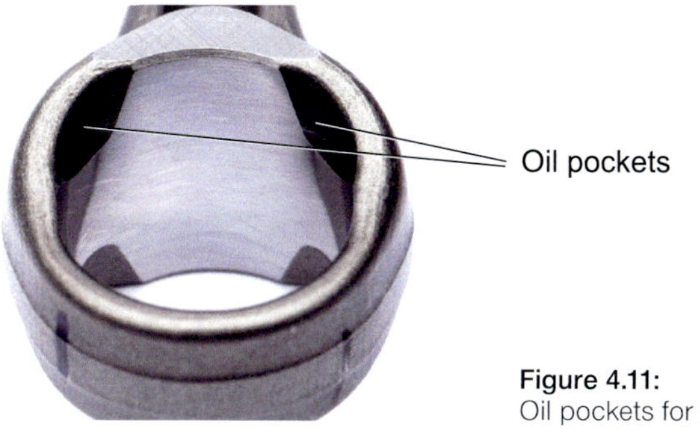


**Figure 4.11:**
Oil pockets for improving lubrication

- A manganese-phosphate layer acts as a protective coating for the run-in phase. The peaks resulting from surface roughness are covered, thus preventing seizure from occurring. As the service life increases, the bore smoothes out and the durability of the running surfaces is ensured.

## 4.7 Guiding the connecting rod

The connecting rod is typically guided axially (in the direction of the crankshaft) by the big end bore on the crank web. This is referred to as "bottom-guided" (**Figure 4.12**).

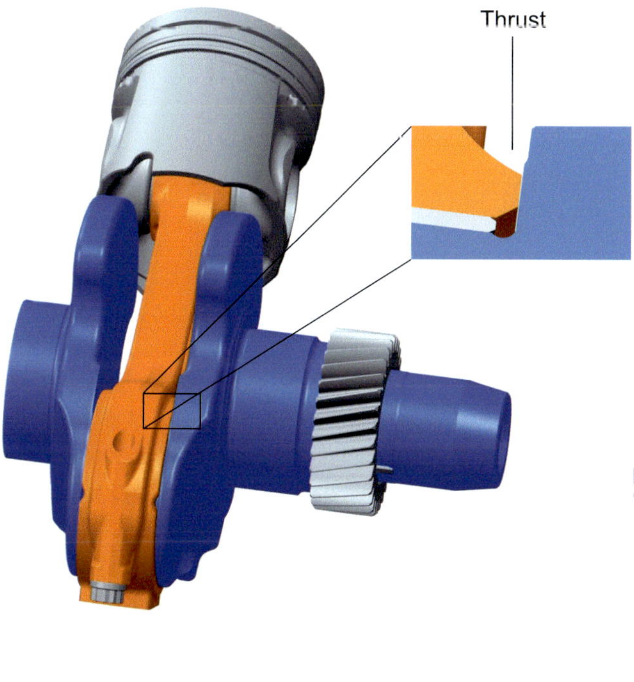


**Figure 4.12:**
"Bottom-guided" connecting rod

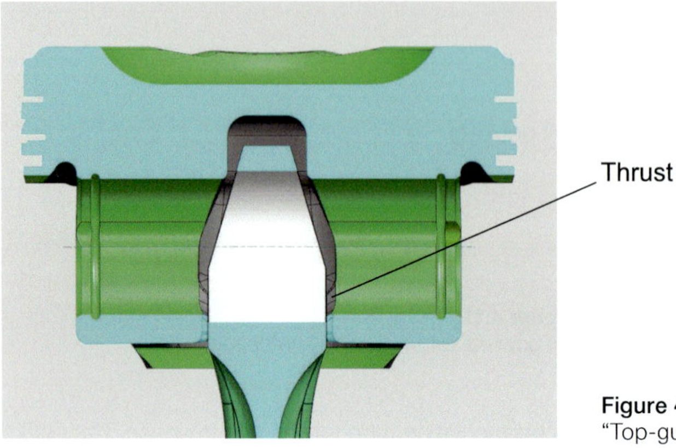


**Figure 4.13:**
"Top-guided" connecting rod

In some cases, the guidance is provided by the piston and the small end bore ("top-guided"); Figure 4.13. To this end, the piston is provided with a surface against which the small end bore runs and the clearance is reduced to about 25 μm.

As a result, the need for thrust washers on the crankshaft is eliminated and the length of the piston pin can be reduced. This leads to a reduction in weight relative to the "bottom-guided" variant. Because of the smaller thrust surfaces, lower friction losses occur.

The disadvantage is that motions and vibrations from the crankshaft are transmitted directly into the piston. The lubrication conditions in the small bore are also worsened, because less oil travels through the narrow gap between the connecting rod and the piston.

## 4.8 FE analysis of the connecting rod

### 4.8.1 Modeling

The starting point of every FE analysis is modeling, i.e., the partitioning of the affected structure into many volume elements. The FE model includes, in addition to the big end bore with the bearing cap, the bearing shells, bolts, and the piston pin, as well as a suitable replacement model for the piston and the crankshaft (**Figure 4.14**). Modeling of all individual components is realized as a three-dimensional structure including all significant details, with only minor simplifications (e.g., screw threads). Symmetrical models can be used to limit the modeling effort for connecting rods for in-line engines. For V-type engine connecting rods, it depends on the number of asymmetries present and is determined for individual cases. The assembled structure is fixed for the analysis solely by means of contact boundary conditions.

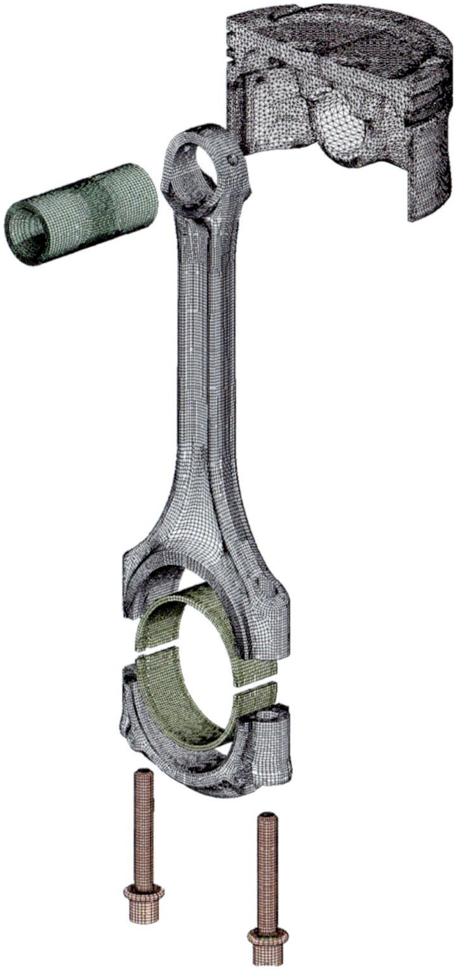

**Figure 4.14:**
Three-dimensional FE model of the connecting rod of a passenger car gasoline engine with bolts, bearing shells, and piston pin, as well as a substitute piston model for load application

Direct securing in position of the connecting rod structure is avoided because it would lead to overconstraint of the conditions at the restraint points. The assignment of material characteristic values concludes the modeling process.

## 4.8.2 Stresses from assembly

The first load case is bolt pretensioning, which results from the assembly of the bearing cap to the connecting rod shank. A prerequisite for realistic determination of the resulting stresses is the consideration of the geometry of the bolt shank, the joint face shape (cracked or machined joint face) and the centering of the joint face, the bolt contact face, thread depth (number of load-bearing thread turns), and bearing crush.

#### 4.8.2.1 Bolt force

Analogous to the specification for tightening the connecting rod bolts, the load on the bolt joint is prescribed for the assembly simulation. In an iterative process, the extension and thus the stress in the bolt shaft is varied until the prescribed pretensioning force of the connection has been reached. The yield point of the bolt material generally limits the amount of pretensioning force. In some cases, however, the contact pressure on the bearing cap in the area of the bolt contact face can be the limiting factor.

As a reaction to the bolt force, the big end bore deforms into an out-of-round blind bore (**Figure 4.15**).

In the connecting rod manufacturing process, a round blind bore is generated by finish machining in the bolted state. It ensures ideal geometry for this highly stressed bearing point. This manufacturing process is also recreated in the simulation by a suitable procedure. This is necessary in order to prevent any prohibited stress increases in the contact zones in the subsequent steps of representing the operating loads.

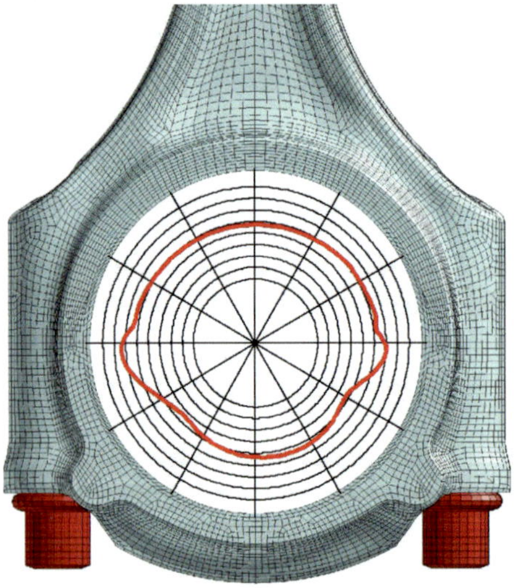

**Figure 4.15:**
Representation of deformation of the blind bore of the big end after connecting rod bolts have been tightened

#### 4.8.2.2 Bushings, bearings, and shrink fit

For the typical plain bearing of the big end bore on the crankshaft, split bearing shells with defined crush height are used for positioning and securing during operation. The bushing in the small end bore, or shrink-fit piston pin in the small end bore, generates stresses due to overlap. This overlap in the bearing shells, piston pins, or bushings is represented in the simulation by appropriate contact boundary conditions. The resulting static stresses are later combined with the dynamic stresses from operating load.

### 4.8.3 Stresses from engine operation

It follows from the kinematics of the crank mechanism that the piston, together with the small end bore and the piston pin, performs an oscillating motion, and the big end bore with the crankshaft pin on the crankshaft primarily performs a rotational motion. The displacement of the big end bore leads to a pivoting motion of the connecting rod. The measure of the pivot angle of the connecting rod is determined by the geometric dimensions of the crank mechanism (crank radius and length of connecting rod).

The pivot motion of the connecting rod leads to alternating transverse acceleration of both the big and small end bores, with an approximately sinusoidal curve (**Figure 4.16**). The lift motion of the connecting rod leads to a longitudinal acceleration, which also features a modified sinusoidal curve. The stroke-connecting rod ratio of the crank mechanism (crank radius to length of connecting rod) determines the degree of deviation from the sinusoidal curve and leads to the acceleration at the top dead center being greater than that at the bottom dead center. The two accelerations would be equal only in the case of an infinitely long connecting rod.

In order to translate the dynamic operating loads on the connecting rod into suitable boundary conditions for a static structural analysis, different load cases that can occur during one or two crankshaft revolutions (depending on the working principle, two- or four-stroke) are captured and applied to the structure in the form of quasi-static boundary conditions.

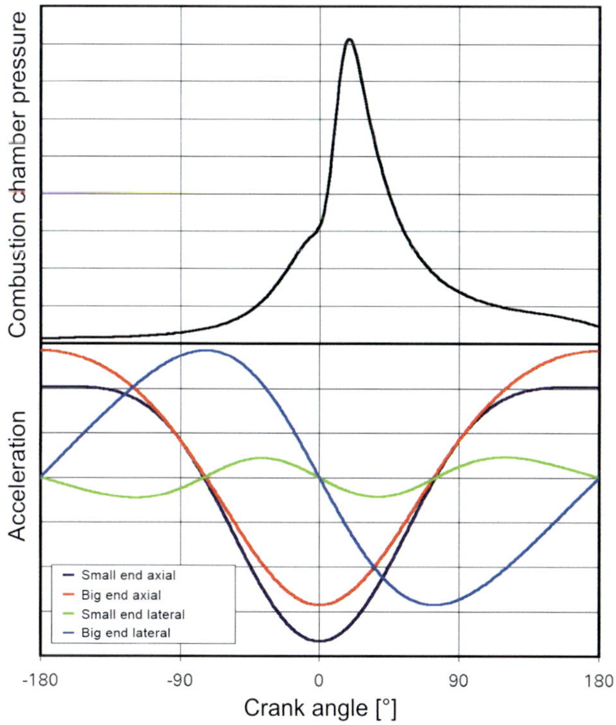


**Figure 4.16:** Curve of acceleration due to gas and inertia forces for a passenger car gasoline engine in a four-stroke combustion cycle

- For the simulation of the load at TDC fired (top dead center in the expansion stroke, the maximum of the combustion chamber pressure is generally applied, so that a slight displacement of the gas force maximum can occur, compared with the representation in Figure 4.16. The relief of the structure due to the inertia force directed opposite the gas force is taken into consideration.
- Only the inertia force, without any combustion chamber pressure load, is therefore used accordingly at the TDC nonfired.
- In order to calculate the load due to transverse acceleration, the respective maxima from the transverse acceleration curve for both the small end bore and the big end bore are applied in combination with the effective combustion chamber pressure at the corresponding point in time.

The individual loads mentioned are combined appropriately to obtain a complete representation of the operating load. Ten relevant load cases are the result. Depending on the engine type, there are different weightings of the individual load cases:

- For passenger car diesel engines, the gas force load on the connecting rod dominates, whereas the load due to transverse acceleration is very small relative to the gas force load and can be neglected.
- For passenger car gasoline engines, likewise, the gas force load on the connecting rod dominates, while the load due to transverse acceleration is small for the typical speed range (up to about 8,000 rpm) and therefore negligible in general.
- For high-speed sport and racing engines, the inertia forces are critical and the loads due to longitudinal and transverse acceleration are correspondingly high. Particularly at very high speeds, the inertia forces can exceed the load due to gas force, and the greatest load magnitude can result, for example, from the transverse acceleration.

#### 4.8.3.1 Gas force

An example of the resulting comparative stresses on the connecting rod structure of a passenger car gasoline engine, under combined assembly, gas, and inertia force loads at the design speed at TDC fired, is shown in **Figure 4.17**.

High stress can be detected on the connecting rod. The location with the minimum cross section establishes the limits of pure compressive capacity. Analyses of the buckling resistance of the connecting rod shank must also be carried out, as the maximum load carrying capacity can be further reduced if the buckling resistance is not sufficient. Modern combustion processes in gasoline engines increase the risk of premature ignition and knocking, which leads to a brief, substantial increase in the pressure load on the connecting rod. These pressure peaks must be considered when computing resistance to buckling. Other locations with high static loading are the areas of bolt force application. The local material creeping that typically results in the bolt thread and in the area of the bolt contact face leads to redistribution and smoothing of the load. Owing to the pulsating gas force load, the transitions from the connecting rod shank to the big and small end bores are dynamically highly loaded locations. The limits of the operational strength of the connecting rod, in terms of service life, ultimately result from this consideration.

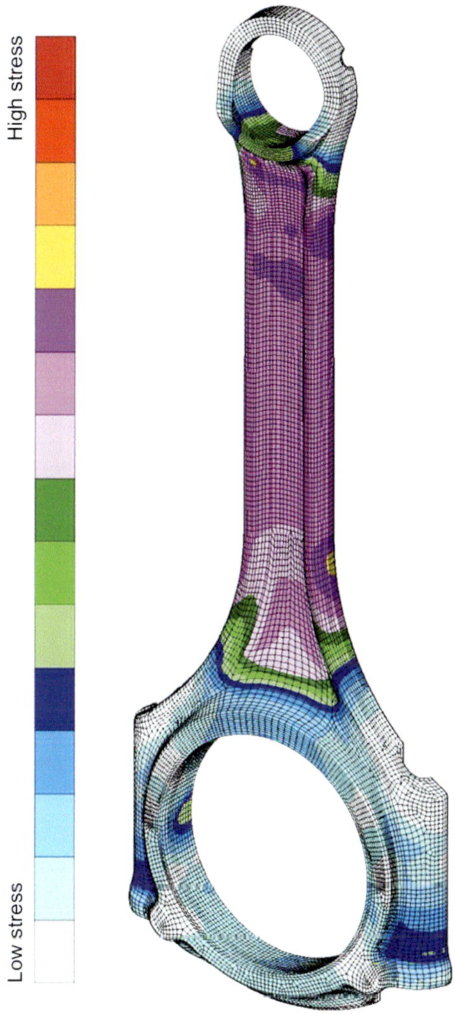


**Figure 4.17:**
Comparative stresses on the connecting rod structure under combined assembly, gas, and inertia force loads at the design speed at TDC fired

### 4.8.3.2 Inertial force

The comparative stresses on the connecting rod structure of a passenger car gasoline engine, under combined assembly and inertia force loads at the design speed at TDC non-fired, is shown in **Figure 4.18**.

Dynamically highly loaded locations from alternating inertial force loads, once again, are the transitions from the connecting rod shank to the big and small end bores. The effect of the inertia force at the TDC nonfired leads to an oval deformation of the conrod bores. The resulting bending load must be borne by the structure at the small end bore and by the screw joint at the big end bore. In addition to the requirements in terms of operational strength, the effects on bearing clearance play a primary role. To limit deformations, larger cross sections may be required than would otherwise be necessary for strength reasons to ensure fatigue resistance. Greater bearing eccentricity may also be called for (cf. Section 3.3.4).

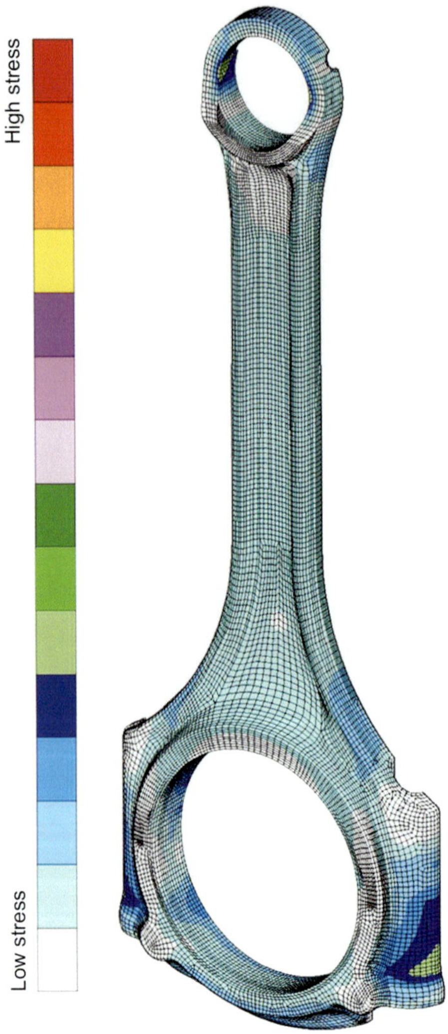


**Figure 4.18:**
Comparative stresses on the connecting rod structure under combined assembly and inertia force loads at design speed at the TDC nonfired

As a further aspect, dynamic gaping of the joint face at the big end bore must be investigated for minimum screw force and maximum inertia force in the TDC nonfired. Gaping must not occur, or has to be minimal, in edge areas, i.e., contact pressure must not reach zero. See **Figure 4.19**. Otherwise, suitable measures to increase contact pressure are needed, such as greater bolt pretensioning forces or a reduced joint face surface area.

**Figure 4.20** shows the comparative stresses in the connecting rod structure of a series passenger car gasoline engine under combined assembly, gas, and inertia force loads at the design speed, at the point of maximum transverse acceleration of the big end bore. This results in only minor bending loads on the connecting rod shank. Bending loads on the connecting rod shank resulting from the maximum transverse acceleration at the small end bore are also low.

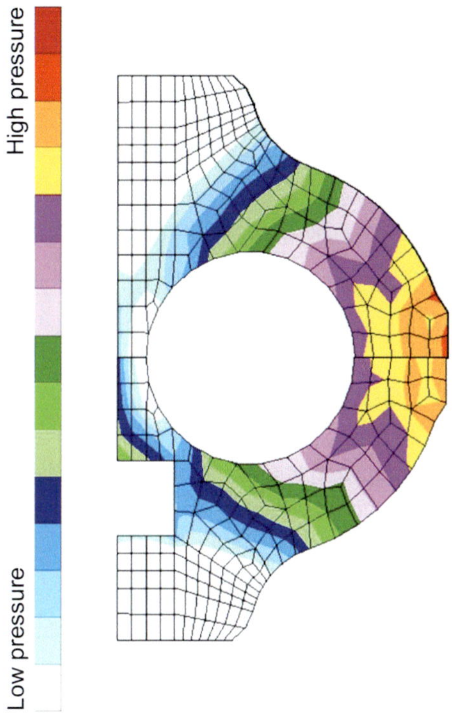


**Figure 4.19:**
Contact pressure distribution for investigation of gaping in the joint face of the big end bore at maximum inertia force loading at the TDC nonfired

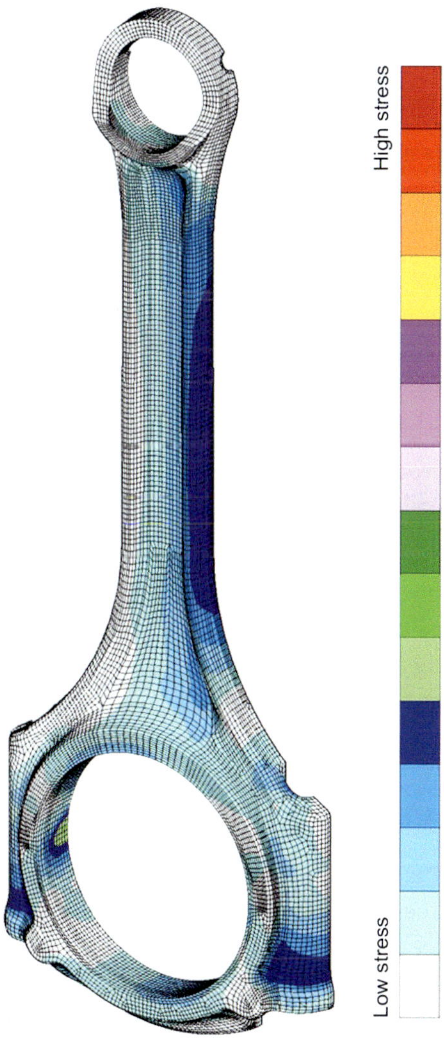


**Figure 4.20:**
Comparative stresses in the connecting rod structure under combined assembly, gas, and inertia force loads at design speed at the point of maximum transverse acceleration of the big end bore

Since the maximum transverse acceleration at the small end bore occurs in an early crank angle range, near the TDC fired and TDC nonfired, the longitudinal load on the structure due to gas force dominates once again. It should be noted, however, that these statements apply exclusively to normal operating speeds for mass-production engines (up to about 8,000 rpm for gasoline and 5,000 rpm for diesel), but not for the very high speeds of racing engines (up to about 20,000 rpm). The greater the rpm level, the more dominant the loads due to inertia forces come to be, until finally they become the largest operating load on the connecting rod.

## 4.9 Component testing of the connecting rod

Connecting rods, like all other components of a combustion engine, must reliably bear the highest loads that can occur during operation. These occur over the entire rpm range when operating under full load.

Testing of component and operational strength is intended to demonstrate that, even considering variability in material strength and the manufacturing process (raw forging, machining, surface treatment, assembly, etc.), the component meets all strength requirements.

The primary stress in tension and compression due to the oscillating inertia force and gas pressure occurs in the axial direction of the connecting rod shank (rod force $F_{St}$). For high-rpm engines, the bending stresses occurring in the plane of motion of the connecting rod, arising from the rotating masses, must not be neglected.

Additional, not insignificant bending moments can act on the connecting rod as a result of production tolerances, installation conditions, and deformations of the crankshaft and the housing.

Typically, component testing of the connecting rod under alternating tension/compression loads is performed on resonance pulsators or servohydraulic testing machines. When determining the load horizon, it must be taken into consideration that owing to mass distribution over the length of the connecting rod, different load conditions R (R = underload/overload) and thus different average stresses can arise during engine operation (**Figure 4.21**).

When determining boundary conditions, execution, evaluation, and statistical evaluation of component tests of this type, various engine manufacturers have specific procedures, based on longstanding experience.

The maximum value occurring during the expansion stroke is always used for the gas force. The inertia force has different values, depending on the engine speed. At MAHLE, the pulsator tests are always carried out using the highest resulting values. This means that the load amplitudes applied in the pulsator test are made up of values that would not necessarily occur at the same speed.

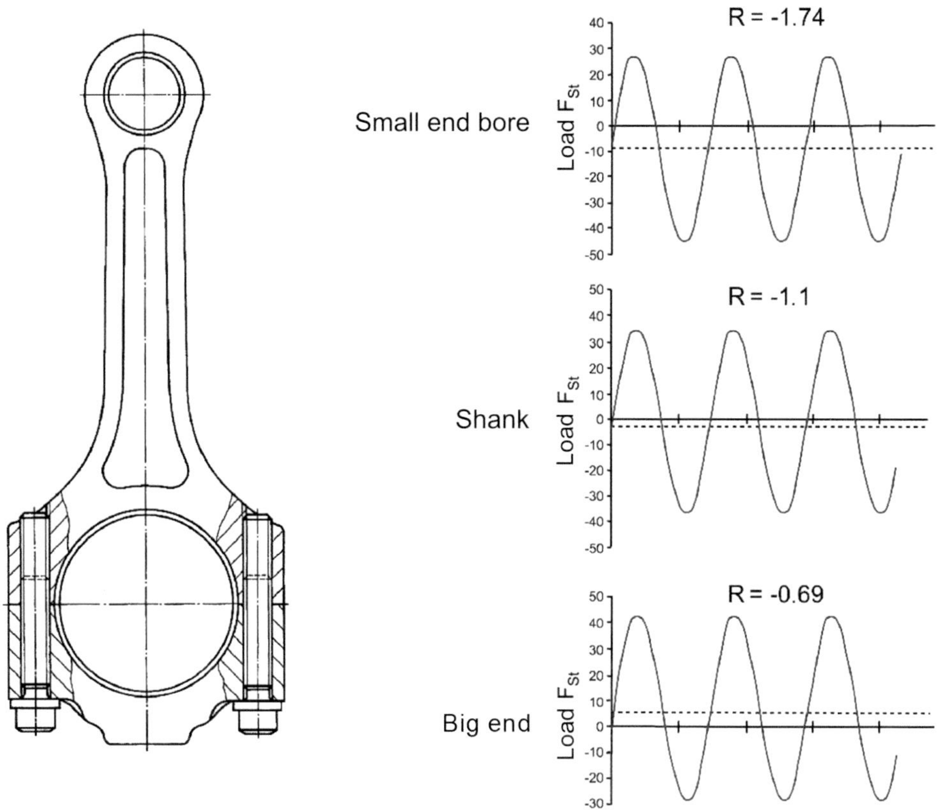


**Figure 4.21:** Pulsator testing of a connecting rod (R: load ratio; $F_{St}$: rod force)

The required data are determined in that, analogous to **Figure 4.22**, the curve of gas force (blue) and inertia force (red) is calculated for different engine speeds, in this case for a four-stroke engine. The resulting sum curve (green), which represents the rod force ($F_{St}$) curve, yields maxima and minima, which are marked here as points 1 and 2.

These are shown in the diagram in **Figure 4.23** as a function of the engine speed. It shows the curve for maximum inertia force (upper curve) and maximum gas force (lower curve), where points 3 and 4 indicate the largest value $F_{St\,max}$ and the smallest value $F_{St\,min}$ of the rod force $F_{St}$.

The following applies for the pulsator tests:

$$F_a = 0.5 \cdot (F_{Stmax} - F_{Stmin}) \qquad \text{Load amplitude} \qquad (4\text{-}1)$$

$$F_m = 0.5 \cdot (F_{Stmax} + F_{Stmin}) \qquad \text{Mean load} \qquad (4\text{-}2)$$

$$R = \frac{F_{Stmin}}{F_{Stmax}} \qquad \text{Load ratio} \qquad (4\text{-}3)$$

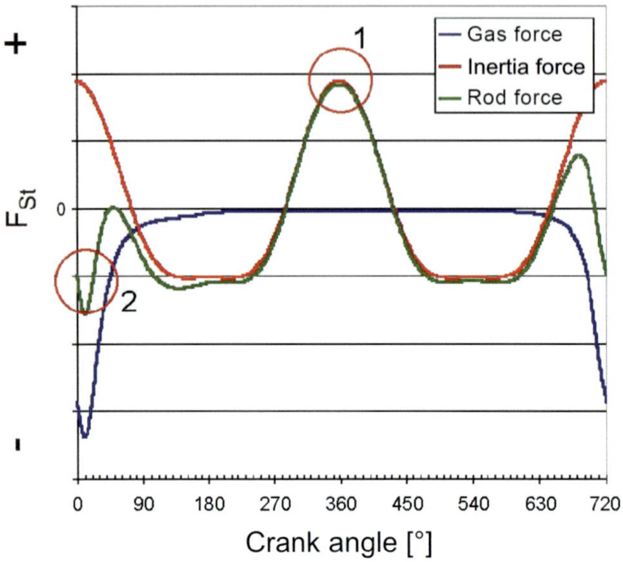


**Figure 4.22:**
Calculated rod forces in a connecting rod over an operating cycle of 720 degrees of angle at constant speed, such as $n_1$

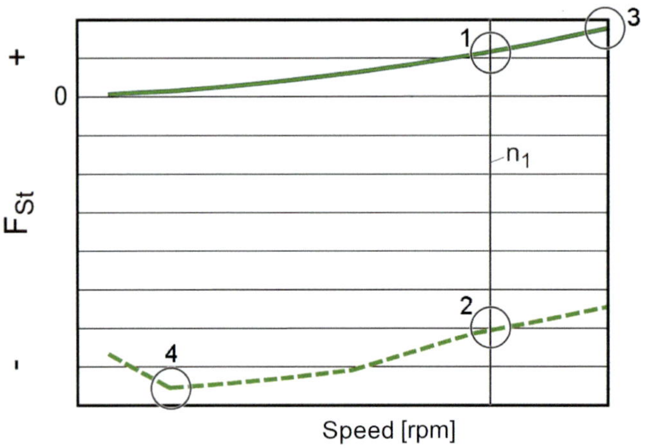


**Figure 4.23:**
Rod forces in a connecting rod as a function of engine speed

The staircase method has been tried and tested for rapid and cost-effective determination of the Wöhler line (**Figure 4.24**). Using a load amplitude near the expected median of the transition range, the first sample is tested. If the sample does not fracture, then the load for the next samples is increased with a constant step width, until fracture occurs. The load is then reduced stepwise until no further fracture occurs. The method very quickly centers on the average value.

A clear combination of the load map in the engine, with lab results of component fatigue resistance, is shown in the Haigh diagram (**Figure 4.25**).

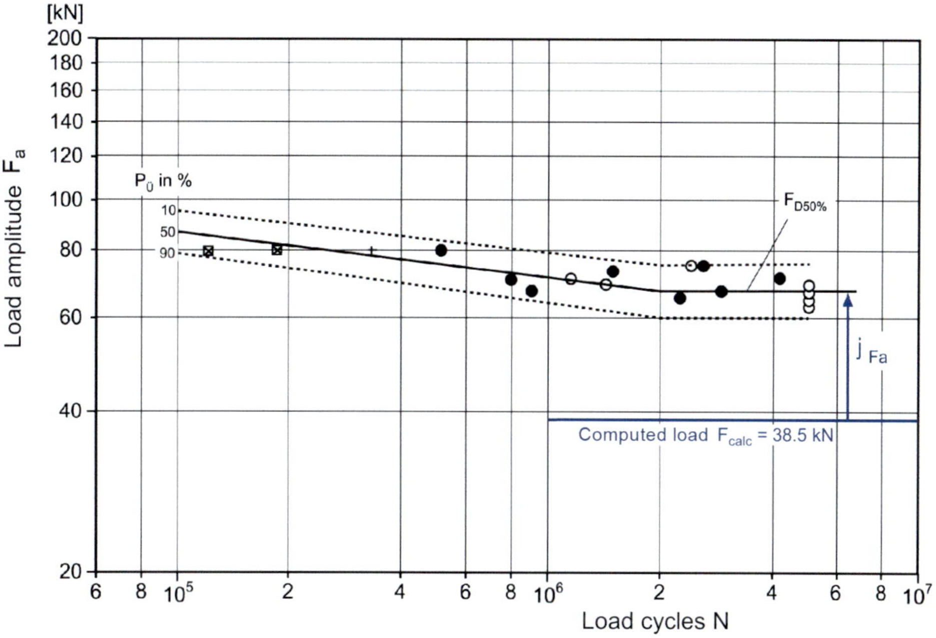


**Figure 4.24:** Component Wöhler lines for a connecting rod, determined according to the staircase method ($P_Ü$: probability of survival at 10, 50, 90%, $F_{D50\%}$: average service life, $P_Ü$: 50%)

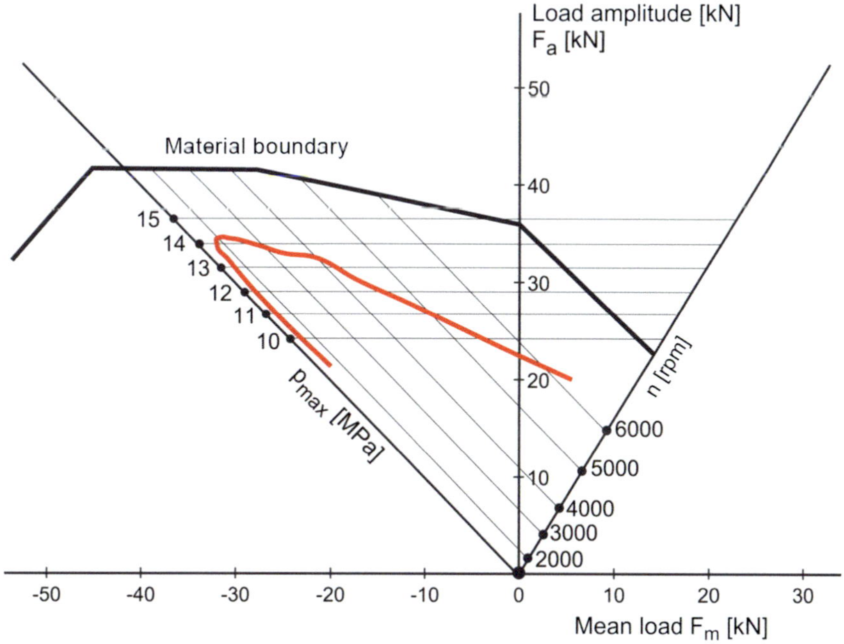


**Figure 4.25:** Operating loads on a connecting rod in a Haigh diagram

The safety factor j is determined from the quotient of

$$j = \frac{\text{Fatigue resistance}}{\text{Operating load}} \tag{4-4}$$

The required minimum value is dependent, among other things, on the required survival probability, the spread, the targeted probability of failure, and the field of application.

In addition to the pulsator tests already described, special fixtures are used to apply, for example, bending loads. Motored tests are also carried out at excessive speeds on a stripped engine, using an external drive.

# 4.10 Materials

## 4.10.1 Steels for forged connecting rods

In the past, heat-treated steels were primarily used for connecting rods. In the 1980s, these were increasingly replaced with precipitation-hardened ferritic-perlitic steels.

With the development of the fracture splitting method (Section 4.4.1), the steel C70S6BY was introduced in the mid-90s as a series material for connecting rods. This material is air-cooled immediately after hot forging, like a precipitation-hardened ferritic-perlitic steel, and has the typical advantages of these steels, such as the elimination of additional heat treatment, low distortion, cost-effective machinability, and good fracture splitting ability. The structure is nearly perlitic, with a small ferrite fraction at grain boundaries. The high perlite fraction supports the formation of brittle fracture surfaces during cracking, which provide an exact fit between the upper and lower parts thanks to their crystalline fracture microstructure.

Increasing combustion pressures and the demand for weight reduction call for the use of higher-strength steel grades, while the requirements for fracture splitting must still be met. The material 46MnVS6mod provides a higher yield point, without degrading fracture splitting ability, which enables an increase of approximately 25% in the design strength of components compared with C70S6BY. These improved mechanical properties have been achieved by increasing the nitrogen and vanadium content and thus the associated precipitation hardening that is typical of precipitation-hardened ferritic-perlitic steels. It is characterized by the formation of finely distributed vanadium-carbonitride precipitation during cooling of the forging from the heat of deformation. The microstructure is ferritic-perlitic and the ferrite fraction is greater than that of C70S6BY, owing to the lower carbon content. Machinability is improved in comparison with C70S6BY.

For the material C70S6BY, increasing the vanadium content can increase precipitation hardening. This material is referred to as 70MnVS4. The increase in strength affects the yield point and the tensile strength equally. Because the tensile strength should not exceed 1,200 MPa for reasons of machinability, the increase in strength is limited to this value.

### 4.10.2 Sinter-forged connecting rods

As an alternative to forged steel connecting rods, connecting rods can be produced in a powder-metallurgical process. In this case, a downstream forging process is used to reduce porosity (sinter forging process). The material used, 3Cu6C, has a carbon content of 0.50–0.60%. The increase in strength is achieved by alloying copper at up to 3.25%. With respect to fatigue resistance, this material is at the level of 70MnVS4. For process-related reasons, the use of sinter-forged connecting rods is limited to passenger car engines. Forged steel connecting rods are used for highly loaded gasoline engines and diesel connecting rods.

## 4.11 Connecting rod bolting

### 4.11.1 Requirements for connecting rod bolting

The connection of the conrod cap to the connecting rod shank is a typical example of a dynamically and eccentrically loaded bolt joint. It transfers inertia forces from the piston, piston rings, piston pin, and connecting rod to the crankshaft pin on the crankshaft. In the process, the forces must be guided around the crankshaft pin. Therefore, in addition to axial loads, transverse forces and bending moments act on the bolt joint. Additionally, as a result of gas forces in the combustion chamber, deformations occur in the big end bore, which causes additional transverse forces in the joint face, particularly for connecting rods with an angle-split big end bore. These boundary conditions lead to dynamic stress in the connecting rod bolts in the longitudinal and transverse directions. To reliably support these stresses, high clamping forces are required.

In addition, the bolt joint has to support the forces for fixing the bearing shells. The force required to generate the overlap from the bearing crush must also be considered in the analysis of the pretensioning force of the connecting rod bolts.

Variations in the pretensioning force should be small, because otherwise undesired shape deviations can occur in the connecting rod bearing. The stress state during machining of the bearing housing, and later during connecting rod assembly in the engine, must therefore be nearly identical, because otherwise the different bolt forces can cause deviations in the roundness of the bearing that negatively affect the function of the bearing.

This makes it necessary to use bolts with high material strength and assembly methods that take as much advantage of the material as possible, up to the yield point, such as the torque/angle method or yield point method. For bolts that are tightened beyond the yield point, the permissible number of times they can be tightened is limited. In some cases, new bolts must be used for repeated assembly.

## 4.11.2 Design and analysis of connecting rod bolting

The design of connecting rod bolts is made on the basis of guideline VDI 2230. It provides general instructions for the analysis of a bolt joint. The derivation of the operating forces on the bolt joint, which result substantially from the inertia force loading due to the masses of the connecting rod and the piston, is not included in this guideline.

Using an analysis method for a closed circular ring model (big end bore), the relevant operating loads (lateral force, transverse force, and bending moment) can be determined in the parting plane of the big end bore (**Figure 4.26**). The calculation of the stresses in the bolt joint uses the maximum tensile load, which is defined in the connecting rod direction by the inertia force at the top dead center.

Starting from the operating load (lateral force, transverse force, bending moment), the systematic calculation steps can be carried out on the basis of guideline VDI 2230. The elastic compliances of the bolts and tensioned parts (stepped bending bodies) are determined and the operationally reliable function of the bolt joint is demonstrated.

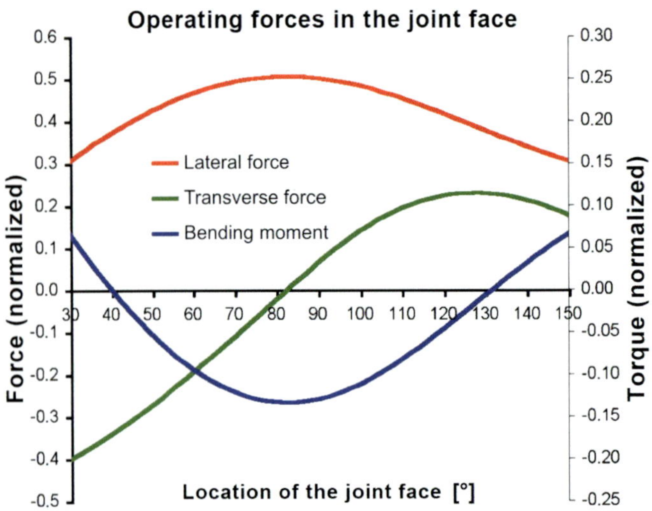


**Figure 4.26:** Calculated forces in the parting plane of the big end bore

The results are

- bolt pretensioning force due to assembly (min., max.) for yield point tightening or rotation angle tightening (superelastic);
- tightening torque (min., max.);
- contact pressure on the bolt contact face;
- required clamping force/bolt pretensioning force to prevent partial gaping of the joint face, considering the clamping force of the bearing shells;
- operating force/engine speed at the start of gaping in the joint face;
- demonstration of durability, stress amplitude (including bending) at the threads, even for the case of partial gaping at the joint face;
- required minimum engagement depth of the thread (nut height).

## 4.11.3 Shape of the connecting rod bolts

The connecting rod bolt joint can be designed as a through hole or tapped blind hole (blind hole with reduced notch effect). The through-bolt with threaded bolts and nuts on both sides, or with a headed bolt and a nut, is mainly used for connecting rods in large-bore engines. In passenger car and commercial vehicle engines, a tapped blind hole using headed bolts is typical.

Connecting rod bolts for passenger car and commercial vehicle engines are designed as bolts with or without a waisted shank, partially or fully threaded, and with or without grooves. Typical head shapes include hex, double-hex, and multitooth, used with external force application (**Figure 4.27**). Bolt strength, $R_m$ ranges from about 800 MPa to over 1,400 MPa in

**Figure 4.27:** Connecting rod bolt design types

exceptional cases. The yield point ratio $R_{p0.2}/R_m$ is greater than 0.9. In order to achieve as great a thread service life as possible, the threads are rolled after heat treatment.

Displacements at the joint face can lead to autonomous loosening of a connecting rod bolt after just a short period of operation, due to settling and wear. They must be avoided without fail. The connecting rod cap and connecting rod have a form-fit connection for reliable fixation. Bolts with a dowel fit or knurl, centering sleeves, pins in flat parting surfaces, toothed parting surfaces, or, as is now typical for passenger car and commercial vehicle connecting rods, fracture parting surfaces (cracking) are used.

In the design of the bolt joint, care must be taken that the bolts are as close as possible to the crankshaft pin. This reduces the risk of gaping in the joint face and reduces bending stress. In angle-split connecting rods, the thread at the higher location in the big end bore is directly in the force flow. Measures must therefore be taken in many cases to increase component strength at the thread exit. The tip of the core hole should have radii and the thread should be rolled in order to further reduce the notch effect.

# 5 Crankcase and cylinder liners

## 5.1 Introduction

The crankcase is the central component of the combustion engine, containing and connecting the functional groups of the crank mechanism, and forms a system boundary that seals off the combustion engine externally. It prevents exit of the working medium, coolant, and lubricant, and the entry of moisture and dirt. The crankcase must make use of the available installation space with the prerequisite of sufficient structural stiffness and with respect to the shape accuracy of the bearing and cylinder bores as well as the cylinder fit (for replaceable cylinder liners), with component mass as low as possible.

Crankcases bear the internal forces and moments and transfer them to the engine mounts. They need to withstand external forces, such as

- forces from add-on parts;
- radial and axial forces from the machine being driven (supporting forces and axial load);
- forces through the engine mounts (e.g. for offroad travel or boat travel with severe waves);
- assembly forces; and
- forces due to thermal expansion.

The type of crankcase is based on the size and application of the engine, the working process (four-stroke or two-stroke), the type of cooling (water/air), the number of cylinders, their design and arrangement, the material, and production process.

A crankcase consists of bearing panels, the side and end walls, cylinder surfaces or liners, and, depending on the design, an upper cover plate. The bearing panels support the crankshaft and, in some commercial vehicle engines, the camshaft(s). It also contains channels for coolant and lubricant ("galleries") and the coolant spaces. The case is closed at the bottom by an oil pan and at the top by the cylinder head. The lower opening entails a loss of structural stiffness for the crankcase. The consequences of this, such as vibrations and deformation, can be compensated for with numerous design measures.

### 5.1.1 Forces and stresses

The gas pressure in the combustion chamber acts both on the cylinder head, which transfers the force via the cylinder head bolts to the crankcase bearing panel, and on the bearing caps via the crankshaft, which are also attached to the bearing panel by bearing cap screws. The force flow is thus closed. The crank case wall is dynamically loaded in tension. The cylinder head bolts are arranged around the cylinders and the cylinder block, and the forces of the bolts in the area of the bearing panel can be guided directly into it. The forces of the bolts in the area of the crank circle plane must be guided to the bearing panel by special design measures, such as tension bands, fins, and belts.

The redirection of the forces into the crankcase bearing panel causes additional stresses. Deformations of the cylinder bore as a result of assembly, thermal, and operational forces can degrade the nominal piston clearance. Deformation of the bores for the crankshaft bearings by the forces in the housing can reach the order of magnitude of the bearing clearance. Detailed CAE analyses (thermomechanical analyses) can support the design process and minimize these problems accordingly.

### 5.1.2 Development goals

In line with fuel consumption and emission goals planned for the future, the power-to-weight ratio of the combustion engine must be minimized (lightweight design concept). Weight is not only a cost factor, but it also affects fuel consumption values proportionally, with corresponding effects on emissions.

Comparing the density of cast iron, at about 7.3 g/cm$^3$, with that of aluminum alloys, at about 2.7 g/cm$^3$, a mass reduction of about 45–55% results for aluminum crankcases, depending on the construction and integration of accessory equipment. The lower stiffness of aluminum, however, requires an adapted design, which significantly reduces the mass advantage.

## 5.2 Types of crankcases

The following types of crankcases are distinguished, depending on the type of connection of the cylinder liners or cylinder surfaces:

- Open-deck design (**Figure 5.1**). The case can be produced in high-pressure die casting.
- Closed-deck design (**Figure 5.2**). This design requires complex sand cores for the water jacket and can be produced in gravity die casting or low-pressure die casting for light alloy designs. The closed-deck design is a more compact and stiffer construction.

In addition to these designs, the types are divided into the so-called skirted block (**Figure 5.3**), where the side walls (skirts) are drawn downward over the main bearing bridge, and the two-piece design, with an upper crankcase and a lower crankcase (also called bearing traverse or bed plate). See **Figure 5.4**).

To prevent deformation of the cylinder surfaces, special design measures are necessary, which also applies to cylinder tubes cast together in the direction of the crankshaft, known as the Siamese variant.

Figure 5.1: Crankcase with open-deck design

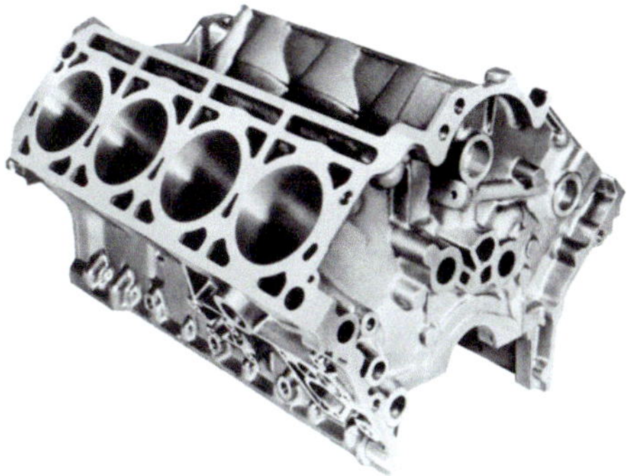

Figure 5.2: Crankcase with closed-deck design

Figure 5.3: Crankcase as skirted block

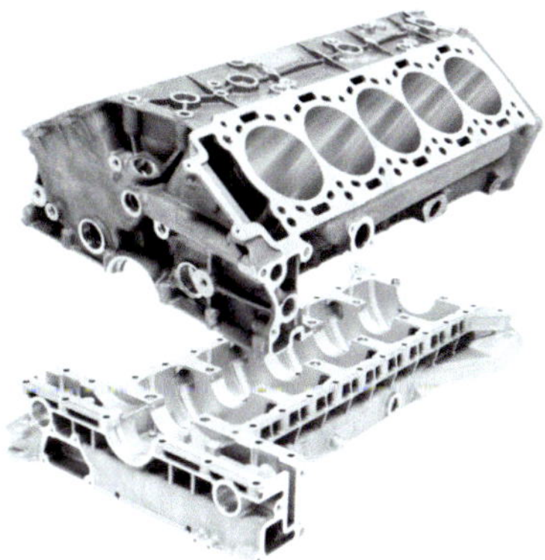

Figure 5.4: Crankcase with two-piece design block and bed plate

## 5.2.1 Methods for attenuating noise emissions

The crankcase is both a source and a transmitter of noise and vibration. The sources include

- broadband combustion noises;
- piston noises;
- vibration excitement of the crank mechanism and valve train;
- natural vibrations of add-on parts;
- natural vibrations of the unit.

These noises and vibrations are transferred through the structure of the crankcase. Excitation of the exterior surfaces causes noise to be emitted. The excitation is also transmitted to the vehicle structure via the engine mounts.

To reduce this vibration and noise emission, larger flat surfaces must be avoided or stiffened by appropriate ribbing, and the bending and torsional stiffness of the crankcase must be optimized. This applies particularly to aluminum crankcases, which must either be ribbed (**Figure 5.5, left**) or come in a two-piece design (**Figure 5.4**). Oil pans and stiffening plates contribute to the bending and torsional stiffness of the overall structure. Transmission mounting flanges and engine mounting locations are also stiffened using ribs.

Cast iron crankcases have a significant advantage over those made of aluminum (better acoustic behavior, lower distortion), due to their higher density, higher Young's modulus, and better damping properties. Cast iron crankcases exhibit less ribbing for this reason (**Figure 5.5, right**).

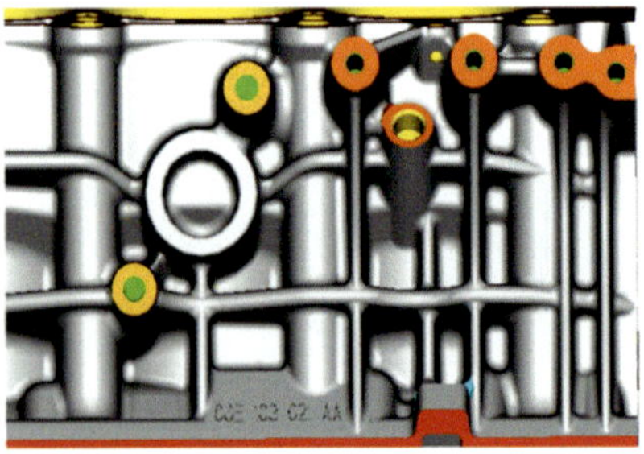

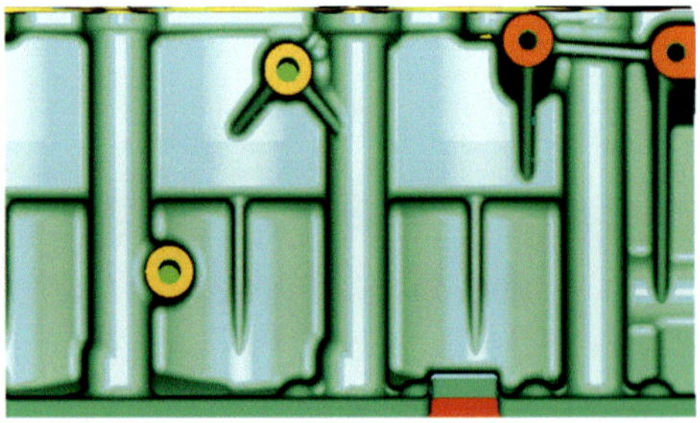

**Figure 5.5:** Segments of crankcases, made of aluminum material on the left, of ferrous material on the right

## 5.2.2 Main bearing seats

The main bearing seats are subject to particularly high loads within the crankcase. When creating the design layout, care must be taken to ensure that no local stress peaks occur (**Figure 5.6**). The bearing seats typically contain threaded holes for mounting the main bearing cap, holes for bay to bay breather, and oil grooves and galleries.

In high-performance engines, the use of several main bearing bolts is a common way to accommodate the high forces from the crank mechanism that are transferred into the bearing cap. For deep skirt crankcases (with crankcase walls that extend very far down), transverse bolts can also be used to further improve the bearing cap structure.

The bearing panel stiffness can be increased with an additional stiffening plate installed between the oil sump connection area of the cylinder block and the bearing cap itself.

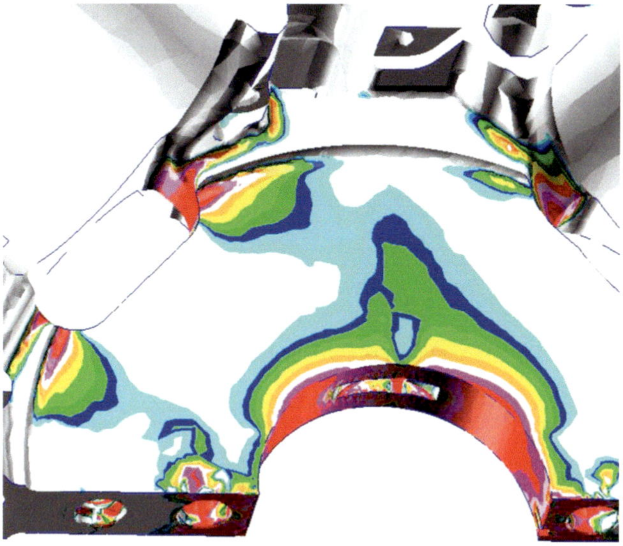

**Figure 5.6:**
Stress curves at a main bearing seat

## 5.2.3 Cooling

The temperatures in the engines must be kept within certain limits for various reasons:

- High temperature gradients cause thermal stresses, which reduce service life.
- High temperatures reduce fatigue resistance in aluminum alloys.
- High temperatures cause large deformations in the crankcase, especially in the area of the cylinder surfaces.
- The cylinder surfaces must be cooled in order to minimize cylinder distortion and overheating of the lubricating oil, especially in the TDC area of the first compression ring.
- Higher temperatures of the cylinder surfaces can make it necessary to shift the ignition point and thus reduce the thermal efficiency.

The operating temperature of the engine is controlled by means of the coolant. It is especially important that the cylinder surfaces have as even a flow as possible on the outside, to prevent thermal deformation or overheating.

The flow of the coolant depends on the design of the cooling system. Normally, the coolant is fed from the exhaust side into the crankcase and from here to the cylinder head.

Today, the cooling jacket is designed using CFD analysis software (computational fluid dynamics, flow simulation) in several optimization cycles. One of the goals is to reduce the amount of coolant needed, so that the engine can reach its operating temperature quickly. Improvements in emission behavior and fuel consumption can be obtained through the use of this process.

Special attention has to be given to the risk of cavitation with cylinder liners, which are in direct contact with the coolant. MAHLE has investigated this problem in extensive development work.

## 5.3 Crankcase materials

Crankcases are cast in iron, aluminum, or magnesium materials. Depending on the application goal, various alloys are available.

### 5.3.1 Cast iron

The most important cast iron materials are GJL (gray cast iron), GJV (cast iron with vermicular graphite), and GJS (cast iron with nodular graphite).

Crankcases made of GJL are

- low cost;
- stable with regard to deformations in both the cylinder surfaces and the main bearings;
- easily machinable.

The material GJL can also be used as a cylinder surface and supports noise dampening. Disadvantages are greater density, lower thermal conductivity relative to aluminum, and lower load carrying capacity compared with GJV and GJS.

GJV has a higher load carrying capacity than GJL, but owing to its severely reduced sulfur content (manganese sulfate acts as a lubricant during machining), it is significantly less machinable than GJL. Because of its higher cost, GJV is currently used only in turbocharged diesel engines with special requirements profiles. If the structure is carefully optimized, GJV also allows thinner walls than GJL. This means that substantially lighter components can be produced.

GJS has a greater load carrying capacity than GJV (tensile strengths of up to 900 MPa). Its disadvantages, however, are higher cost, reduced castability, and poor thermal conductivity.

### 5.3.2 Aluminum alloys and material properties

Aluminum alloys stand out thanks to a combination of good thermal conductivity, low weight (**Table 5.1**), easy machinability, and acceptable mechanical properties. An advantage of aluminum crankcases with aluminum cylinder surfaces is that the installation clearance between

**Table 5.1:** Weight savings with aluminum, compared with GJV, using the example of a V6 crankcase

| Crankcase | GJV<br>[%] | Aluminum low-pressure die casting<br>[%] | Aluminum sand casting<br>[%] | Aluminum sand casting with iron cylinder liners<br>[%] |
|---|---|---|---|---|
| Cast | 100 | 54 | 42 | 50 |
| Machined | 100 | 52 | 42 | 45 |
| Completed | 100 | 58 | 50 | 55 |
| Completed, with oil pan | 100 | 60 | 50 | 55 |

the piston and cylinder surface can be smaller than if cast iron liners are used, because of the similar thermal expansion coefficients. Piston noise is reduced at the same time.

The weight advantage and better thermal conductivity improve the thermal efficiency and thus the fuel consumption and exhaust gas emissions.

These advantages are countered by lower stiffness, higher material and process costs, and reduced strength values of aluminum, especially at temperatures of greater than 200°C, as is shown in **Figure 5.7**.

The most important alloying elements for the use of aluminum in crankcases are magnesium, manganese, copper, and silicon. Manganese, magnesium, and copper are typical constituents for improving the mechanical strength of aluminum. Particularly above 150°C, copper improves the strength properties of aluminum-silicon alloys.

Silicon improves casting properties and wear behavior of the microstructure of the cylinder surfaces. In crankcases made of hypereutectic alloys (Si ≥ 12.5%), a minimum land width of 4 mm may be achieved between cylinders.

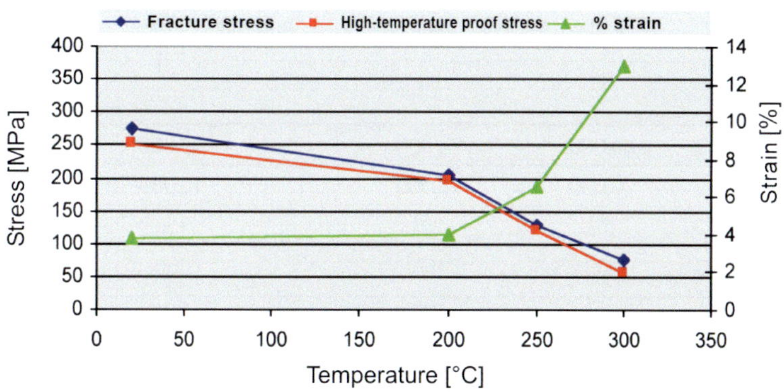


**Figure 5.7:** Sample strength values of an aluminum alloy as a function of temperature

**Table 5.2:** Chemical composition of aluminum alloys used at MAHLE for cylinder liners

| Alloy symbol | | **MAHLE 147** AlSi17Cu4Mg | **MAHLE 233** AlSi10MgCu | **MAHLE 124V** **MAHLE 124P** AlSi12MgCuNi | **226 (EN-AC 46000)** GD-AlSi9Cu3 (Fe) as per EN 1706 |
|---|---|---|---|---|---|
| Alloying elements, % by weight | Si | 16.0–18.0 | 9.0–11.0 | 11.0–13.0 | 8.0–11.0 |
| | Cu | 4.0–5.0 | 0.6–1.0 | 0.8–1.5 | 2.0–4.0 |
| | Mg | 0.4–0.7 | 0.2–0.5 | 0.8–1.3 | 0.05–0.55 |
| | Ni | – | max. 0.15 | 0.8–1.3 | max. 0.55 |
| | Fe | max. 0.7 | max. 0.6 | max. 0.7 | max. 1.3 |
| | Mn | max. 0.2 | 0.1–0.4 | max. 0.3 | max. 0.55 |
| | Ti | max. 0.2 | max. 0.15 | max. 0.2 | max. 0.25 |
| | Zn | max. 0.2 | max. 0.3 | max. 0.3 | max. 1.2 |
| | Cr | max. 0.05 | – | max. 0.05 | max. 0.15 |
| | Al | remainder | remainder | remainder | remainder |

**Table 5.3:** Mechanical and physical properties of MAHLE aluminum alloys

| Alloy symbol | | **MAHLE 147** | **MAHLE 233** | **MAHLE 124V** | **MAHLE 124P** |
|---|---|---|---|---|---|
| Manufacturing process | | LP permanent mold casting | LP permanent mold casting | LP permanent mold casting | forged |
| Hardness HB10 | | 90–120 | 85–110 | 90–125 | 90–125 |
| Tensile strength $R_m$ [MPa] | 20°C | 180–220 | 190–250 | 210–230 | 300 |
| | 150°C | 160–210 | 180–220 | 180–200 | 250 |
| Yield point $R_{p0.2}$ [MPa] | 20°C | 160–210 | 160–210 | 190–210 | 280 |
| | 150°C | 150–190 | 150–200 | 170–180 | 230 |
| Elongation at fracture A5 [%] | 20°C | 0.5 | < 1 | < 1 | < 1 |
| | 150°C | 0.5 | 1 | 1 | 4 |
| Fatigue strength under reversed bending stress $\sigma_{bw}$ [MPa] | 20°C | 70 | 80 | 90–100 | 130 |
| | 150°C | 60 | 70 | 75–80 | 115 |
| Young's modulus [MPa] | 20°C | 84,000 | 78,000 | 80,000 | |
| | 150°C | 79,000 | 76,000 | 77,000 | |
| Thermal conductivity $\lambda$ [W/mK] | 20°C | 152 | 155 | 155 | |
| | 150°C | 153 | 156 | 156 | |
| Thermal expansion [$10^{-6}$ m/mK] | 20–100°C | 19.4 | 22 | 19.6 | |
| | 20–200°C | 20.4 | 22.4 | 20.6 | |
| Density $\rho$ [g/cm$^3$] | 20°C | 2.7 | 2.7 | 2.68 | |

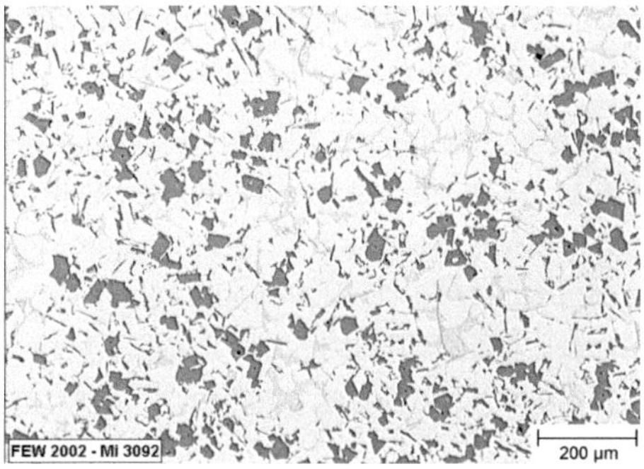

a) MAHLE 147

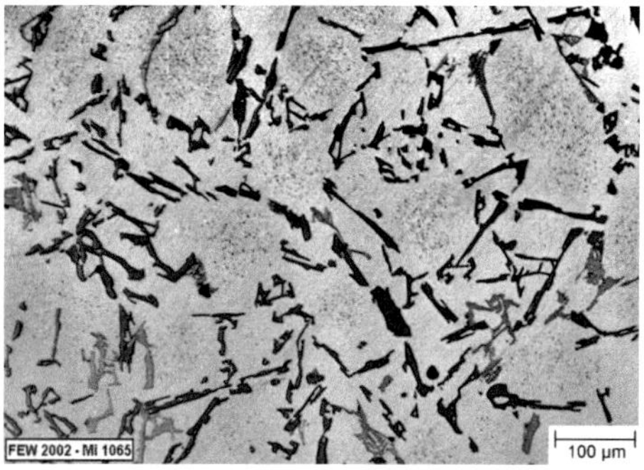

b) MAHLE 233

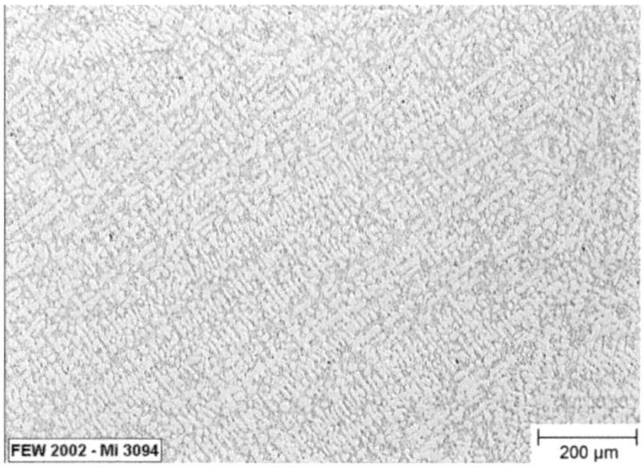

c) MAHLE 124V

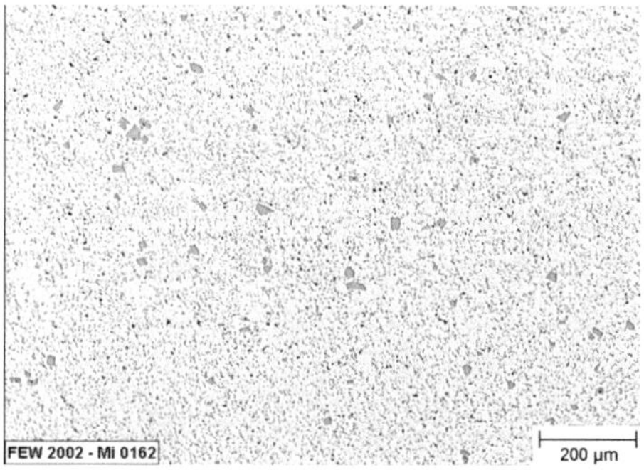

d) MAHLE 124P

**Figure 5.8:** Microstructure images of MAHLE aluminum alloys

**Tables 5.2** and **5.3** give an overview of the compositions and properties of the aluminum alloys used at MAHLE for crankcases, cylinder heads, and cylinder liners. Typical material microstructures are presented in **Figure 5.8**.

Among Al cylinder alloys, the hypereutectic AlSi alloy MAHLE 147 (comparable to the standardized US alloy Reynolds 390) takes a special position. Owing to its high content of primarily precipitated hard silicon crystals (**Figure 5.8a**), it features outstanding wear resistance. Combined with special machining of the cylinder surface, which exposes the silicon crystals, good running properties can be obtained even without coatings. The alloy MAHLE 147 is particularly well suited for low-pressure permanent mold casting and is widely employed for monolithic crankcases. It can also be used to make cylinder liners.

The hard silicon crystals, however, present a challenge for machining. For this reason, crankcases with running surface coatings or cylinder liners are preferably made of more easily machinable, hypoeutectic AlSi alloys. As a high-strength material that is also well suited for processing in sand casting or low-pressure permanent mold casting, the alloy MAHLE 233 is used (**Figure 5.8b**). It was developed on the basis of the standardized alloy 233 (EN-AC 43200 per EN 1706) and contains Cu as an additional alloying element to improve strength at elevated temperatures. Bearing traverses (bed plates) for split crankcases can also be made of this alloy. Today, however, it is common to produce the bearing traverses from standard alloys with cast-in iron inserts in order to support the main crank mechanism loads.

For NIKASIL®- and CROMAL®-coated running surfaces on finned cylinders in air-cooled engines, which are subjected to relatively high temperatures, the eutectic alloy MAHLE 124V is used (**Figure 5.8c**). It is a variant of the piston alloy MAHLE 124 with a modified structure. Thanks to the refinement, the mold filling behavior of the alloy is improved, among other things, which is important for casting very thin-walled fins. For less highly stressed finned cylinders made by high-pressure die casting, the standardized alloy 226 (EN-AC 46000 per EN 1706) is employed.

For forged cylinder liners in motorsport engines, the piston alloy MAHLE 124P is used in combination with a NIKASIL® running surface coating (**Figure 5.8d**).

### 5.3.2.1 Effects of the casting process on the material properties of aluminum alloys

The local material properties of a crankcase cast in aluminum depend on the local mold filling velocity, cross sections, and thus the local cooling speed during casting (**Figure 5.9**).

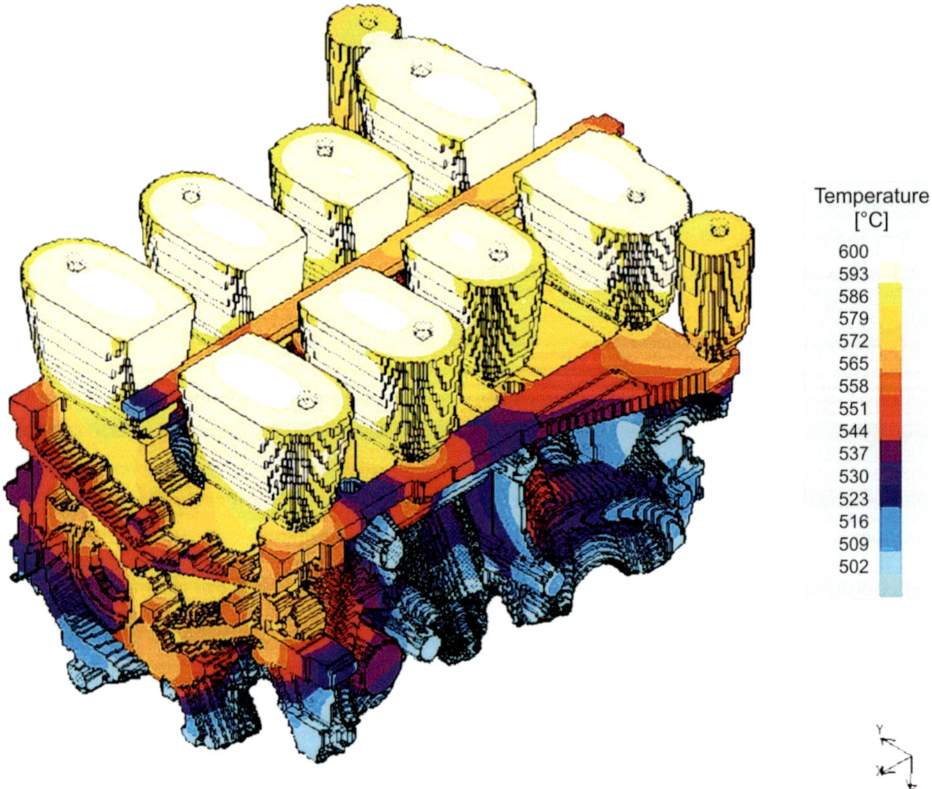


**Figure 5.9:** Analysis of the solidification profile of a crankcase in the mold

Before creating a casting setup, MAHLE performs a casting simulation using a CFD program to determine the following data in particular:

- Defining a filling and solidification profile
- Liquid fraction of the melt
- Failure analysis
- Solidification time
- Dendrite spacing
- Not exceeding a maximum permissible porosity
- Optimization of cooling in the tool

A projection of the residual stresses and local material properties is performed in a further analysis, using a special analysis program.

#### 5.3.2.2 Effects of heat treatment on the properties of cast aluminum alloys

Heat treatment of cast aluminum alloys serves to adjust the material properties. It must be customized for each material and subsequent application.

One of the possibilities is heat age hardening after casting. The soak time and temperature determine the final mechanical properties of the product.

The fatigue strength can be improved by hot isostatic pressing (HIP). In this process, the casting is held in an inert atmosphere at high pressure and temperature prior to heat treatment. The material is thus compacted and porosity is reduced.

For sand castings, heat treatment can consist of a multistage cycle (**Figure 5.10**). During solution heat treatment, the castings are soaked for several hours at a temperature just below the melting point, at which the constituents enter solution. The part is then quenched. This can, however, result in residual stresses in the component. Hot age hardening (artificial aging) follows.

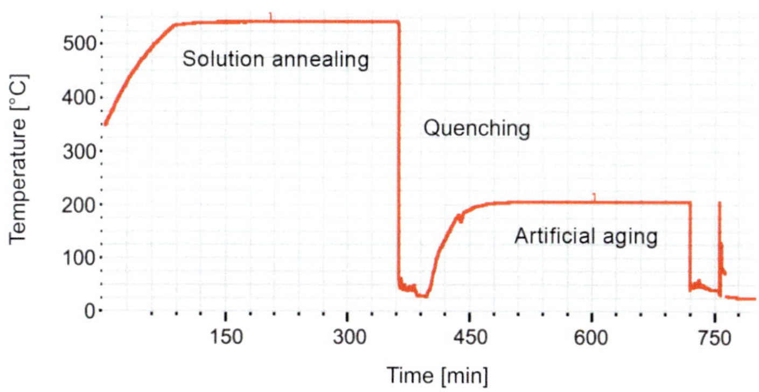


**Figure 5.10:** Sample heat treatment curve of a sand casting made of an aluminum alloy

### 5.3.3 Magnesium

Owing to its density, 35% lower than that of aluminum, and its low Young's modulus, magnesium is typically used for non-structural engine components. As a material for crankcases, it has some disadvantages:

- Increased tendency to creep at high temperatures
- Loss of preload at higher temperatures when using steel bolts
- Tendency toward galvanic corrosion when in contact with steel
- Tendency to corrode when exposed to coolants

One possible solution for reducing component weight is to produce highly stressed locations in a hypereutectic aluminum alloy, which are then encapsulated with magnesium. With a hybridized combination of magnesium and aluminum, weight savings in such a component can be up to 25%.

### 5.3.4 Material trends

The trend toward ever-higher power-to-weight ratios requires new materials. This applies first to passenger car diesel engines with working pressures of up to 200 bar, which constitutes the performance limit for the use of aluminum crankcases. For less stressed engines, such as gasoline engines, aluminum will remain dominant, except for cases where cost is a significant factor or where high-performance derivatives are sought. Magnesium will find only limited application owing to its lower high-temperature stability and tendency to creep, its corrosive behavior, and its higher cost.

### 5.3.5 Effects of the casting process on the design of the crankcase

The technical requirements and tools required, and therefore the costs, have a significant effect on the selection of the casting process. This decision is generally made prior to the first draft of the component. It is therefore important to be aware of the advantages and disadvantages of the individual casting processes. Some of the typical casting processes and their associated designs are shown below.

#### 5.3.5.1 Sand casting

Sand casting parts are cast in a sand mold with sand cores inserted. These can be individual cores or core packages (**Figure 5.11**). The molten metal is either poured from above or pumped into the mold from below. **Figure 5.12** shows a mold with mold filling from below. The cores are generally manufactured in sand or in a CPS process (core package sand process).

**Figure 5.11:** Inserting a core package in a mold

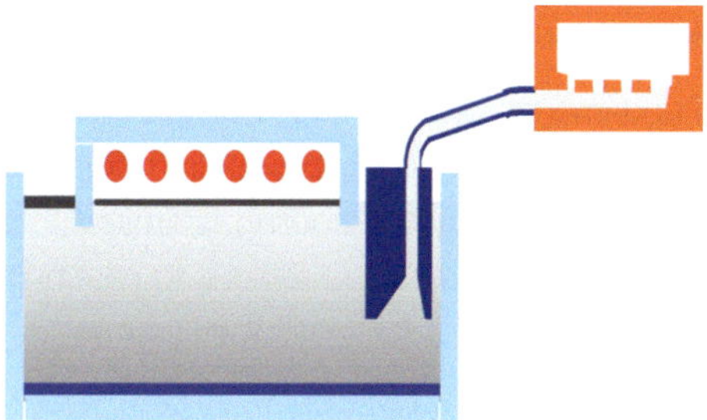

**Figure 5.12:** Principle of a mold with filling from below

### 5.3.5.2 COSCAST™ process

One variant of the sand casting process is the patented COSCASTTM process, in which cores are made from zirconium sand. This material provides precision casting with high surface quality, thanks to its shape accuracy. The density of the sand is very similar to that of aluminum. This means that the casting cores move only very slightly, further improving the dimensional accuracy of the manufactured parts. The aluminum melt is filled into the mold from below using a ceramic pump; it is then rotated 180° for solidification. The process is particularly suited for small and medium batch sizes. The zirconium sand can also be recycled in a thermal process and reused.

### 5.3.5.3 Molding sand–"green sand"

Sand cores made of molding sand are bonded with clay or mud and have a relatively high moisture content. The moisture is absorbed by the aluminum during casting and can cause an unacceptable level of porosity in the part, which is associated with an undesired reduction in mechanical strength. This process is not always suitable for series production, despite its cost advantages.

### 5.3.5.4 CPS method

In this method, the sand core is chemically bonded with resin, for example. The binder is then hardened in a furnace, or by infiltration with gas, or even just at room temperature, depending on the type. Silicon or zirconium sands are used. Zirconium has a very small thermal expansion coefficient and density similar to that of aluminum. Cores made of zirconium sand are used preferably for die-cast parts with high dimensional accuracy, for complex cores, and for high mechanical strength requirements.

The principal advantages of sand casting for crankcases are the possibilities of

- incorporating oil and other galleries as well as cooling jackets in the casting;
- casting bores and recesses (weight savings);
- improving material properties with heat treatment.

Fundamental disadvantages of this method include the high investment cost required for mass production (long production times) and the difficulty of cooling critical regions (the potential for targeted cooling is limited to the use of cooling inserts). The lack of cooling capability makes it more difficult to cast hypereutectic alloys.

#### 5.3.5.5 Full-mold casting method (lost foam method)

Cores are made of polystyrene foam, given a heat-resistant coat, and placed in the mold. When the liquid metal is poured in, the cellular material is gasified.

Complex crankcases with a high number of integrated parts can be manufactured at a reasonable cost using cores of this type. The ability to integrate water-cooled galleries in the mold pack is limited, however, so it can be difficult to reliably achieve the required cooling speed at critical points.

#### 5.3.5.6 Permanent mold casting

In permanent mold casting, reusable molds and cores are used and the metal is cast into the mold under pressure.

The characteristics of this process are high dimensional accuracy, very good surfaces, and high tooling costs. Crankcases are generally manufactured in permanent molds at medium to high production quantities.

#### 5.3.5.7 Gravity die casting

The molten aluminum is filled into the mold under the influence of gravity. Because of the low filling pressure, lost sand cores can even be used, and there is only very low turbulence. Dimensional accuracy is good. The local material structure can be improved, if needed, with targeted cooling. The material properties can also be stabilized with targeted heat treatment, such as solution heat treatment, quenching, and artificial aging (Section 5.3.2.2).

#### 5.3.5.8 Low-pressure die casting

Filling the mold under controlled pressure opens up a few advantages compared with normal gravity die casting. By controlling the filling speed, better mechanical properties can be obtained, and by maintaining pressure during the solidification process, the porosity of the part is controlled in a targeted manner.

#### 5.3.5.9 High-pressure die casting

The aluminum is filled into the mold under high pressure. High pressure and rapid filling allow the casting of thin-walled parts, improvement of material properties with intensive cooling, and short cycle times. The disadvantage is that the use of insert parts is limited and expensive because the risk of absorption of gases, and thus porosity, is increased, and subsequent heat treatment is not possible without degassing the mold. It is typical to use cast iron inserts for cylinder liners or bearing inserts in aluminum die-cast components.

#### 5.3.5.10 Squeeze casting

Compared with "simple" high-pressure die casting, as described in Section 5.3.5.9, the melt is filled into the mold through a rise at low speed. High pressure is maintained during the solidification process with this method as well. Crankcases are preferably cast vertically in this process, which allows effective degassing during the filling process and thus reduces gas inclusions. Subsequent heat treatment is used to obtain improved material properties. The required wall thicknesses for this process are generally somewhat greater than for high-pressure die casting. The use of an open silicon matrix structure as an insert can also improve the local material properties (e.g., in the cylinder region of the crankcase).

#### 5.3.5.11 Semisolid process

A newly developed casting process for manufacturing crankcases processes aluminum in a semisolid (thixotropic/rheotropic) state. In this process, the aluminum is heated to the appropriate temperature and then injected into the mold under high pressure. For thixotropic casting, the aluminum billets with non-dendritic structures are inductively heated to the appropriate temperature and then injected into the mold. For rheotropic casting, the liquid metal is cooled to the semiliquid phase, in which no dendrites have yet formed, and then injected into the mold. Both processes prevent entrapped air, feature low shrinkage and short solidification times, form fine-grained structures, and provide the capability of improving material properties through further heat treatment.

## 5.4 Cylinder liners and cylinder surfaces

### 5.4.1 Requirements for the cylinder surface

Because of the lift motion of the piston and piston rings, their partner, the cylinder surface, is also subjected to wear. Wear occurs particularly at the top dead center of the piston ring because the change in direction of the moving parts limits the lubrication. The wear behavior of the running surface and the piston rings is substantially determined by the material pairing selected for the two components. In order to reduce wear, the running surface should be smooth and the lubrication between the sliding partners must be ensured. The type and quality of the running surface affect oil consumption as well as the wear of the two components.

## 5.4.2 Cylinder surfaces in aluminum crankcases

The monolithic aluminum crankcase is based on a hypereutectic AlSi alloy (such as AlSi 17), where the cylinder surfaces are produced by chemical etching or mechanical exposure (special honing) of the primary silicon crystals. The disintegration rate of the silicon crystals must not exceed an upper threshold for both processes.

For the quasi-monolithic crankcase, the running surface is coated, for example with a galvanic layer of MAHLE NIKASIL® (**Figure 5.27**), by means of a plasma-thermal spray process (Plasma Transfer Wire Arc (PTWA)), High Velocity Oxygen Fuel (HVOF), or using twin arc spraying. Alternatively, running surfaces can be produced with local material engineering, by laser alloying (e.g., with silicon) or by using Al matrix composite materials (preforms) with subsequent finishing.

The dominant construction for automotive applications, however, is the heterogeneous crankcase, in which cast-in or inserted cylinder liners made of GJL form the running surfaces.

In addition to the general requirements for cylinder surfaces, additional conditions must be met by cylinder liners. The wall thickness and material strength must be sufficient, so that the cylinder liners do not crack. The finite element analysis allows the construction and material selection to be adapted to the loads due to assembly, temperature, peak cylinder pressure, and piston lateral forces. Stresses originating in assembly are essentially determined by the number, tightening torque, and arrangement of cylinder head bolts as well as the selected cylinder head gasket. **Figure 5.13** shows a typical gas pressure and lateral force curve as a function of the crank angle. The maximum lateral force is later in time than the maximum gas pressure, while the lateral force acts transverse to the pin axis only in the piston contact area.

The loads on the cylinder liners, however, like the stresses and deformations at the circumference, are different because of the temperature distribution, lateral forces, and bolt arrangement. In **Figure 5.14**, using the example of the bottom side of the flange of a cylinder liner, the resultant stresses for the load case of assembly and temperature and the superposition of all load cases at maximum lateral force are depicted. The maximum stress occurs in the area of the radius of the cylinder liner contact with the crankcase. Using fatigue resistance charts, the effects of changes in construction and material on the local safety factor are evaluated.

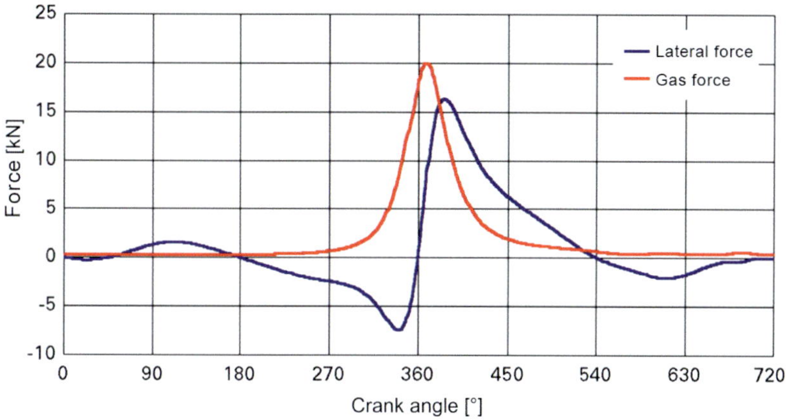


**Figure 5.13:** Gas force and lateral force as a function of the crank angle

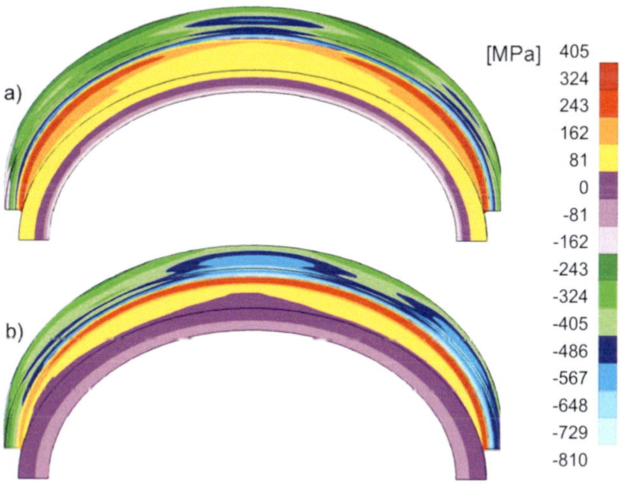


**Figure 5.14:** Maximum stresses at the bottom side of the flange of a cylinder liner, for the following load cases:
a) Assembly + temperature + max. lateral force + gas force
b) Assembly + temperature

## 5.4.3 Types of cylinder liners

The design of the cylinder liner is based on the field of application of the engine, among other things.

For passenger car engines, cylinder liners are cast in the aluminum crankcase. Replaceability of the cylinder liner does not seem to be necessary because of the relatively low service life of these engines.

a) b)

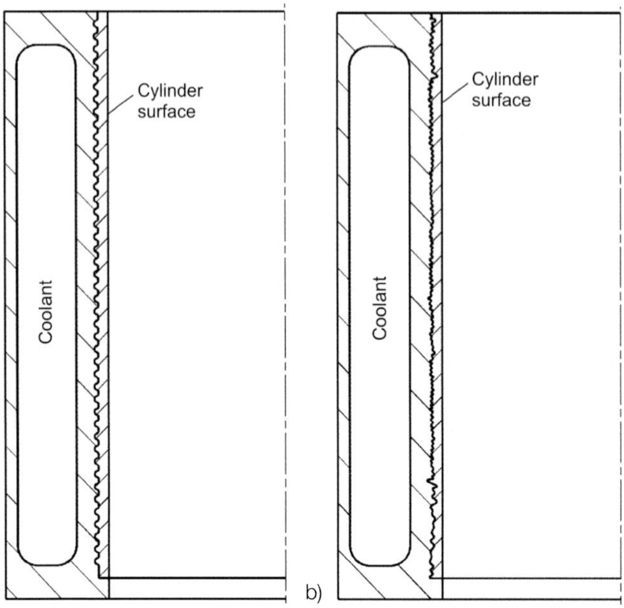


**Figure 5.15:**
Cast-in liners for light-alloy housing:
a) Threaded cylinder liner
b) Rough cast liner

To obtain a form-fit connection during casting, the outer surface of the cylinder liner is threaded, such as by machining (**Figure 5.15a**). So-called rough cast liners are cast in with a rough surface (**Figure 5.15b**). Rough cast liners provide excellent heat dissipation from the combustion chamber due to their strong bond with the crankcase. A light alloy coating on the external surface of these cylinder liners is also possible in order to improve bonding to the case material.

The geometric design of the cylinder liners has to be adapted to the installation space design of the engine and to the casting process. The cylinder liners are fine bored and honed in the crankcase (Section 5.4.5). The remaining load-bearing residual wall thickness for rough cast liners made of gray cast iron should be no less than 1–1.5 mm. The wall thickness depends on the permissible land width between the cylinder liners, the roughness depth on the outer diameter, and offsets during casting.

When used in high-pressure die-cast cases, roughness depths of about 1.5 mm can be used for rough cast liners. When casting in a gravity die casting process, the roughness depth should not exceed 1 mm. For the land spacing between two cylinder liners, different limit dimensions apply depending on the casting method (gravity die casting, high-pressure die casting with single or double gating).

In addition to the typical rough cast liners made of cast iron, MAHLE has also developed a rough cast liner made of a hypereutectic aluminum-silicon alloy and an aluminum liner composite.

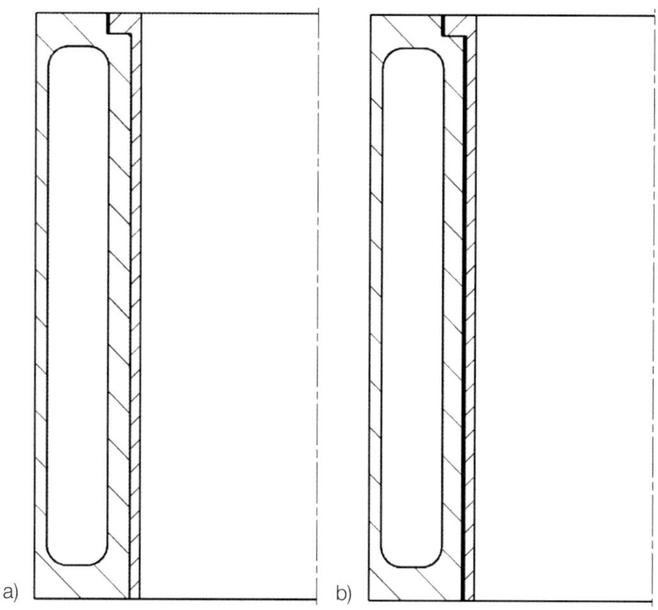


**Figure 5.16:**
Dry cylinder liners:
a) Pressed-in cylinder liner, press fit
b) Inserted cylinder liner, transition fit

In the case of large-bore engines for power generation or engines for commercial vehicles, the running surface wears relatively severely as a result of the long cylinder service life. Because the service life of these engines is generally much greater than for passenger car engines, replaceable cylinder liners made of suitable materials are generally used (Section 5.4.4).

For the cylinder liners used in the crankcase, a differentiation is made between "dry" and "wet" liners.

Dry cylinder liners made of cast iron have relatively thin walls at 1.5–4 mm. This means that they take up only little installation space, but are not in direct contact with the coolant (**Figure 5.16**). Precise matching of installation clearances between the crankcase and the cylinder liner is necessary. This prevents stress peaks that can cause cracks, and reduces the loss of contact between the insert and the crankcase structure.

The axial location of replaceable cylinder liners is determined by a flange, generally on the top side. Dry cylinder liners are mounted with a press or interference fit (**Figure 5.16a**) or a transition fit (e.g., H7/n6) (**Figure 5.16b**) in the crankcase. Cylinder liners with interference fits are finish machined only after being installed in the crankcase.

Cylinder liners with interference fits feature better heat transfer to the crankcase. This solution is selected, for example, when cast iron liners are installed in aluminum crankcases. Good heat transfer is ensured, however, only if the interference exists at all operating temperatures present in the aluminum case. Fits similar to ISO fits N7/r6 (interference ~ 0.05–0.10 mm) can be used for installation here or, for engines with increased operating temperatures, R7/r6

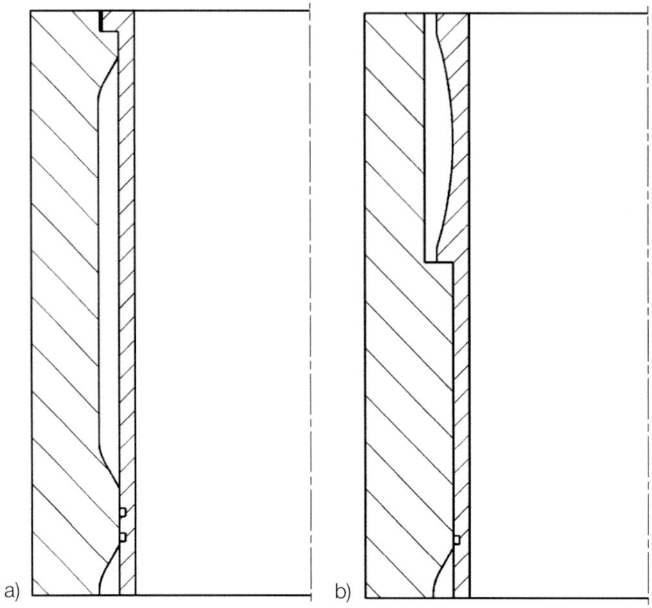


**Figure 5.17:**
Wet cylinder liners:
a) Standard
b) Midstop liner

(interference ~ 0.075–0.125 mm). These interference fits make it necessary, as a rule, to cool the cylinder liners with liquid nitrogen during assembly or to heat the crankcase.

For a cast iron crankcase, the problem of loosening does not exist, because the cylinder liners and crank case have a similar thermal expansion coefficient. An ISO fit K7/r6 (overlap ~ 0.03–0.08 mm) is recommended. In any case, the overlap to be selected must be matched to the operating temperatures of the engine and the respective assembly process.

Cylinder liners with a transition fit, which are already finish machined, are used in commercial vehicle engines with medium performance. Dry cylinder liners, however, are no longer used as a first choice in modern high-performance engines. Wet cylinder liners are used for this application.

Wet cylinder liners are in direct contact with coolant (**Figure 5.17**), thus ensuring excellent heat dissipation. For commercial vehicle and large-bore engines, cylinder liners made of cast iron are preferred. The required wall thickness depends mainly on the maximum pressure in the expansion stroke. With increasing peak cylinder pressures and larger displacements, high-strength materials such as GJS (cast iron with nodular graphite) or steel are increasingly used.

Wet cylinder liners are divided into standard (**Figure 5.17a**) or midstop liners (**Figure 5.17b**). The standard liners feature cooling over the entire running surface area. The midstop liners have a flange within the running surface area. In this design, water is fed only to the upper part of the running surface area, which is thus cooled more efficiently. In both cases, location

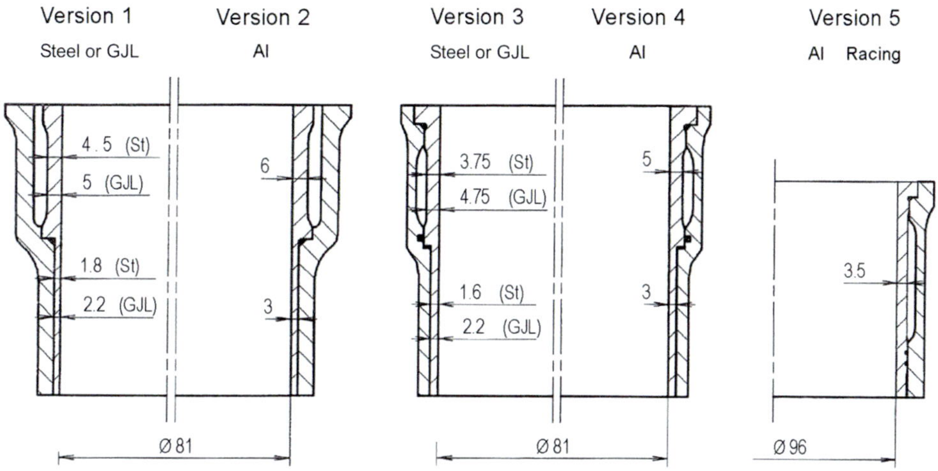


**Figure 5.18:** Comparison of wall thickness ratios of gray cast iron (GJL), aluminum (Al), and steel (St) for passenger car applications (data in mm)

fixing and sealing is done at the top with the cylinder head gasket and the cylinder head. The water chamber is sealed radially at the bottom with O-rings.

In the passenger car sector, sporty high-performance engines still use wet cylinder liners made of steel or coated aluminum.

Steel cylinder liners are thinner, while aluminum ones are somewhat thicker than cast iron liners (**Figure 5.18**).

A special group consists of head-cooled cylinder liners, which are used in large-bore engines. This type of cylinder liner has bores for coolant flow in the head of the cylinder wall that protrudes from the crankcase.

In order to prevent buildup of oil carbon on the top land of the piston in large-bore engines, or at least to make it more difficult, a ring made of cast iron can be inserted in the cylinder liners, loosely located in a turned recess inside the top of the cylinder liner. The ring protrudes slightly into the combustion chamber, thereby scraping the oil carbon off the top land of the piston. For commercial vehicle diesel engines, the use of such a ring is difficult because there exists less available installation space for the ring. In this case, a knurl can be used in the cylinder liner, for example, or scraper rings made of sheet metal are an option.

## 5.4.4 Materials

Cast iron, coated or uncoated aluminum alloys, or steels are used as materials for cylinder liners. Aluminum liners are coated with NIKASIL®, and steel liners can be hardened, reinforced, or coated with CROMAL®.

The requirements for installation space, the operating conditions of the engine, and the cost all determine the selection of material for cylinder liners and their properties. Important properties are specific weight, microstructure, hardness, tensile strength and fatigue strength under vibration, thermal conductivity, thermal expansion coefficient, stiffness, and Young's modulus. Cast iron is preferred for diesel engines. Cylinder liners made of aluminum provide significant advantages in thermal conductivity and specific weight. Steel stands out for its high strength and stiffness.

Cast iron, in many alloy variants, has been a proven and cost-effective cylinder liner solution for decades. See **Table 5.4**. Lamellar gray cast iron (GJL) with perlitic basic microstructure can be manufactured with tensile strengths of up to about 350 MPa. To ensure wear resistance, phosphorus or carbide-forming elements are added, or the running surfaces are inductively hardened. Phosphorus forms hard steadite, which forms as a network if the phosphorus proportion is sufficiently high. For strengths above 400 MPa, lamellar gray cast iron (GJL) with a bainite basic microstructure or cast iron with nodular graphite (GJS) can be used; see **Figures 5.19** to **5.21**.

The alloys listed in **Table 5.2** are used as aluminum-based liner materials. For steel liners that are coated, simple structural steels can be used.

**Table 5.4:** Properties of cast iron for cylinder liners (reference values)

| **Properties** | | **Gray cast iron (GJL)** | | **Cast iron with nodular graphite (GJS)** |
|---|---|---|---|---|
| Basic microstructure | | perlite | bainite and very fine perlite | perlite |
| Hardness [HB] | | 180–300 | 270–330 | 260–330 |
| Tensile strength [MPa] | | 200–350 | 400–600 | ≥600 |
| Young's modulus [GPa] | | 100–120 | 120–140 | ≥150 |
| Chemical composition (weight %) | C | 2.8–3.3 | 2.6–2.8 | 3.1–3.7 |
| | Si | 1.8–2.1 | 1.4–2.0 | 2.1–2.8 |
| | Mn | 0.6–1.0 | max. 0.8 | 0.35–0.65 |
| | P | max. 1.0 | max. 0.08 | max. 0.1 |
| | S | max. 0.12 | max. 0.08 | max. 0.02 |
| | Cr | 0.1–0.3 | – | – |
| | Mo | max. 0.6 | 1.0–1.5 | – |
| | Cu | max. 0.8 | – | max. 0.1 |
| | B | max. 0.07 | – | – |
| | Ti | – | – | – |
| | Ni | max. 1.2 | 1.0–1.5 | max. 1.0 |

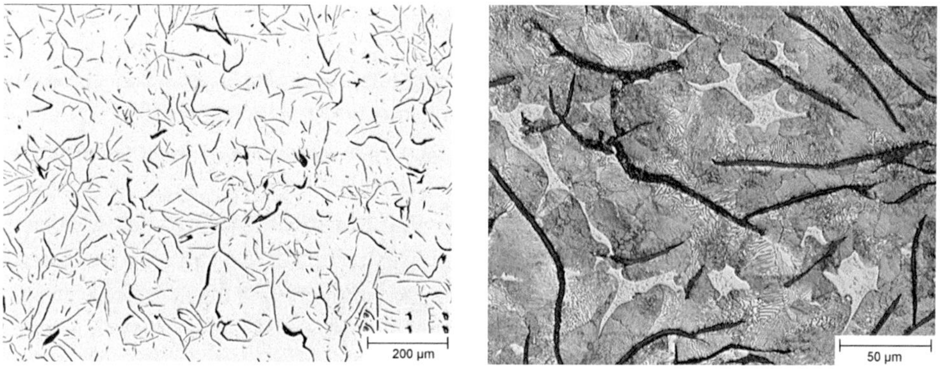


**Figure 5.19:** Gray cast iron (GJL)–perlite
Left: unetched, graphite: type A and B, size 4–6; right: etched, basic microstructure: perlite, max. 5 % ferrite, steadite network

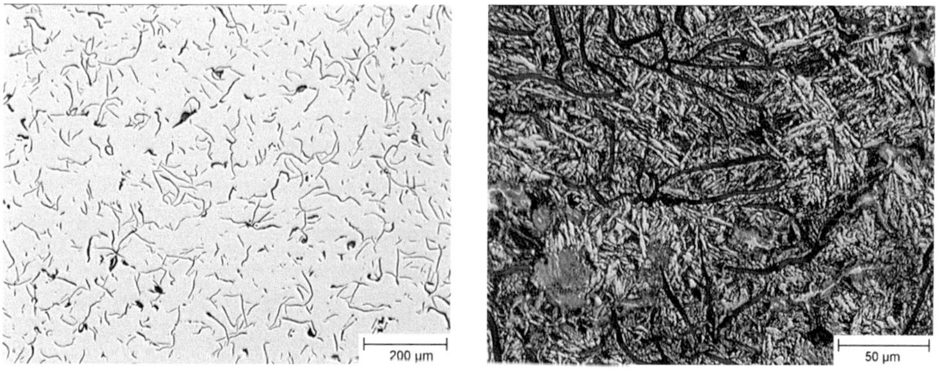


**Figure 5.20:** Cast iron–bainite
Left: unetched, graphite: type A and B, size: 4–6; right: etched, basic microstructure: bainite and very fine perlite

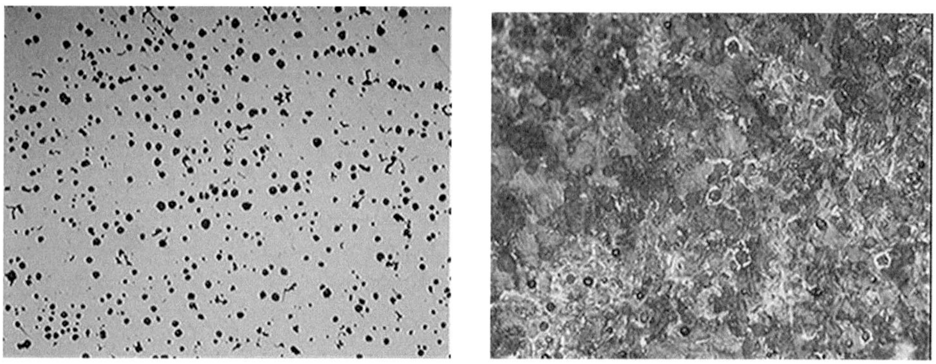

**Figure 5.21:** Cast iron with nodular graphite (GJS)–perlite
Left: unetched, graphite: nodular graphite; right: etched, basic microstructure: perlite with max. 5% ferrite

### 5.4.5 Surface treatment

The running surface structure generated by honing has a very significant influence on the operating conditions of the engine, the running-in characteristics of the piston rings, and the wear of the sliding component. During honing, a cylindrical tool equipped with honing stones performs a rotational and a vertical motion simultaneously. This generates a cross scored structure in the cylinder. Various processes are employed at MAHLE, such as normal honing, brush honing, plateau honing, and sliding honing. The honed surface can be characterized by roughness measurements, fax-film, white light interferometry, and metallographic investigations. To evaluate the roughness measurement of a cylinder surface, the values standardized in DIN EN ISO 13565-2 according to **Table 5.5** are typical.

For hypereutectic aluminum-silicon cylinder liners, as with crankcases, the running surfaces are chemically etched or mechanically exposed. Eutectic AlSi cylinder liners receive a NIKASIL® coating (**Figure 5.27**).

**Table 5.5:** Roughness parameters for honed surfaces

| Parameter | Description | Influence on function |
|---|---|---|
| $R_{pk}$ | Reduced peak height | Many and high peaks above the core roughness depth ($R_k$) extend the run-in phase and can lead to increased wear. |
| $R_k$ | Core roughness depth | The core roughness depth largely determines oil consumption. The smaller the core roughness depth, the lower the oil consumption. The achievable core roughness depth depends on the honing method. |
| $R_{vk}$ | Reduced scoring depth | The deep scoring forms the oil reservoir. An $R_{vk}$ value that is too low can lead to increased long-term oil consumption. |
| $M_{r1}$ | Material ratio of protruding peaks | In order to keep the run-in phase short, the smallest possible value should be targeted. |
| $M_{r2}$ | Material ratio of core roughness depth | Values depend on the honing method. Together with $R_{vk}$, they represent a dimension for oil volume capacity ($V_0$). |

## 5.5 Light-alloy cylinders

The design characteristics of light-alloy cylinder barrels, like those of crankcases, are determined by the individual application.

## 5.5.1 Types of light-alloy cylinders for small engines

For each application case, the following characteristics are distinguished:

- **Cooling**
  Air flow cooling, fan cooling, water cooling
- **Engine design**
  Installed cylinders, cylinder block; cylinders or cylinder blocks form a unit with the crankcase and cylinder head.
- **Operating principle**
  Cylinders for four-stroke engines as installed cylinders with continuous bore
- In **two-stroke engines**, the cylinders are significantly more complicated, because they contain the intake, scavenging, and exhaust ports, and possibly clean air channels required for gas exchange; the vast majority are manufactured with a cast-on cylinder head, and some with crankshaft bearings.

## 5.5.2 Air-cooled cylinders

Air-cooled cylinders are fitted with cooling fins on the external circumference for cooling, to increase the effective heat transfer area. Theoretically, cooling fins with a cross section that gets smaller toward the outside are the most effective. For cast cylinders, however, for manufacturing reasons, these are slightly trapezoidal with rounded edges at nearly the same cooling capacity (**Figure 5.22**).

The heat transfer at the cooling fins increases in case of

- increased fin area (closer spacing, greater fin height);
- increased cooling air velocity;
- transition from free-flowing to guided cooling airflow;
- transition to a cylinder material with higher thermal conductivity (e.g., from gray cast iron to aluminum).

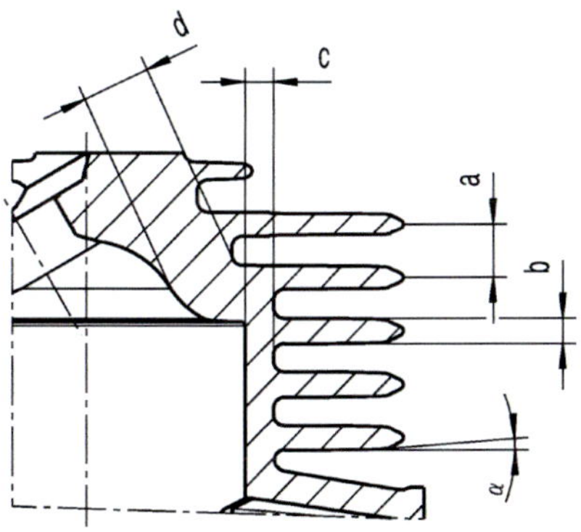


| Casting process | a | b | c | d | α |
|---|---|---|---|---|---|
| Sand casting | 6.5 | 1.4 | 5 | ≥ c+1 | 1° |
| Low-pressure die casting | 6.5 | 1.4 | 6 | ≥ c+1 | 1°10′ |
| High-pressure die casting | 6 | 1 | 5 | 1 | 1° |

**Figure 5.22:** Minimum dimensions for cast cooling fins (minimum values in mm)

The influence of the individual factors is captured rather accurately with appropriate analysis methods. Owing to the very complex relationships of the entire heat exchange process between the cylinder filling and the cooling air, however, theoretical determination of cylinder temperature is very complex. Measurements on the engine are more cost-effective. For air-cooled engines with high specific power output, light-alloy cylinders have been tried and tested, but cooling capacity can be increased only to a limited extent by extending the cooling fins. Fin heights greater than 60–80 mm yield only a slight improvement in heat dissipation, even for light-alloy cylinders. Extending the cooling fins is more effective for light-alloy cylinders than for those made of cast iron. Irrespective of a few exceptions, fin spacing closer than 6 mm in cast cylinders is not practical for casting reasons. **Figure 5.22** contains the minimum dimensions in the fin area that are important for manufacturing. For low fin heights, fin spacing of at least 4 mm can be implemented by machining.

## 5.5.3 Port shapes and gas exchange in two-stroke engines

High specific engine output is required of most small two-stroke engines. High speeds and efficient gas exchange are necessary. Because the time for gas exchange gets shorter and shorter as the speed increases, the control cross sections, and thus the port cross sections, must be as large as possible.

The emissions limits required by law also increase the complexity of the design. Weight plays a large role, particularly for hand-held power equipment. Integral solutions are therefore needed in order to meet these emissions values.

In the conventional two-stroke engine, purging losses occur because of the principles involved. This means that unburned air-oil-fuel mixture is "purged" into the exhaust, which increases fuel consumption and emissions. Air purging in particular represents a now typical solution for minimizing purging losses. An air cushion is built up ahead of the fuel-air mixture through additional clean air channels, so that the purging process sends almost exclusively fresh air into the exhaust gas tract.

The angle and port time area derived from the control diagram in **Figure 5.23** defines the control cross sections effectively available during gas exchange for intake, mixing, and exhaust.

**Figure 5.24** describes the relationship of specific engine output to port time areas, on the basis of 100 $cm^3$ displacement, for various high-performance two-stroke engines. This design of the specific port time areas in the control diagram enables a significant reduction in practical experimentation.

The complexity of the gas exchange and the valve timing has increased with the introduction of the air-purged two-stroke engines mentioned above. Depending on the design of the air purging, additional clean air channels are required.

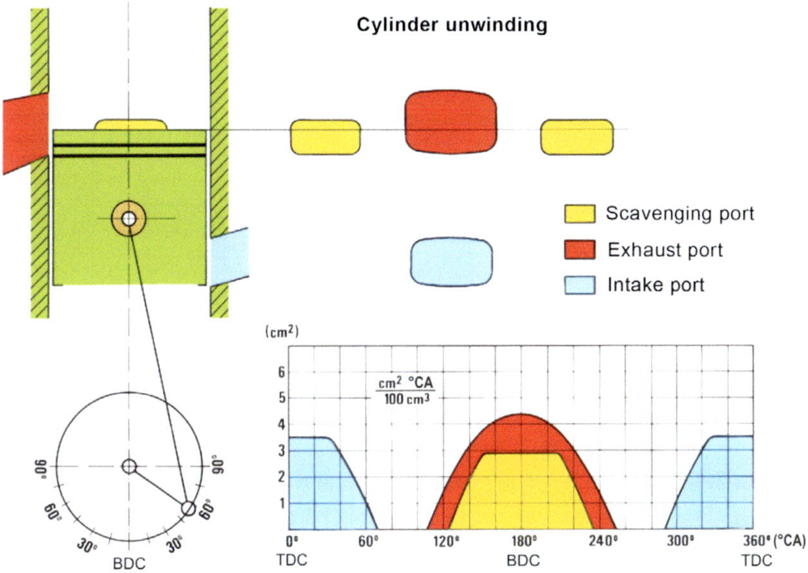


**Figure 5.23:** Time sequence of charge cross sections in two-stroke engines

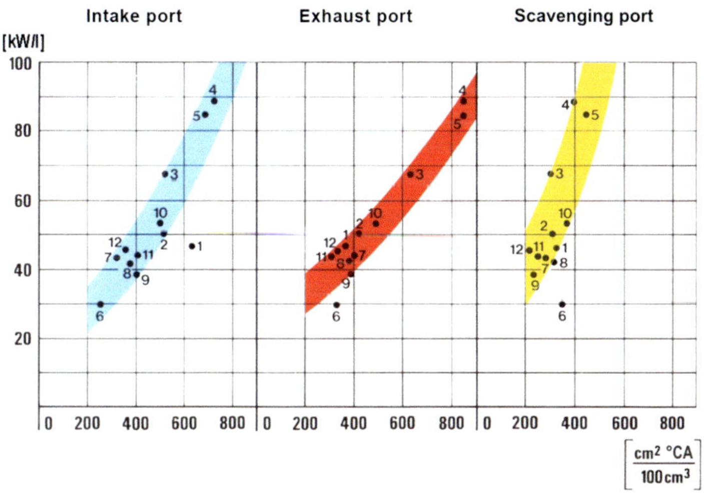


**Figure 5.24:** Examples of specific port time areas in high-performance two-stroke engines

| Serial no. | Application | Displacement |
|---|---|---|
| 1 | Chain saw | 31.7 $cm^2$ |
| 2 | Chain saw | 44.4 $cm^2$ |
| 3 | Motorcycle | 47.6 $cm^2$ |
| 4 | Motorcycle | 49.9 $cm^2$ |
| 5 | Motorcycle | 49.9 $cm^2$ |
| 6 | Motorized bicycle | 49.9 $cm^2$ |

| Serial no. | Application | Displacement |
|---|---|---|
| 7 | Chain saw | 55.5 $cm^2$ |
| 8 | Chain saw | 60.8 $cm^2$ |
| 9 | Chain saw | 68.7 $cm^2$ |
| 10 | Motorcycle | 73.1 $cm^2$ |
| 11 | Chain saw | 80.5 $cm^2$ |
| 12 | Chain saw | 105.5 $cm^2$ |

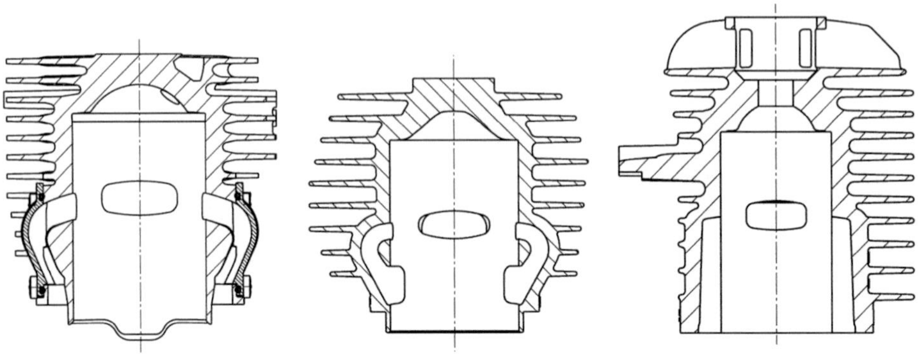

**Figure 5.25:** Design type of scavenging ports in two-stroke engines

If the channel ports in the cylinder surface are too wide, the running behavior of the piston rings is degraded and the shape stability of the cylinder is negatively affected. The width of the exhaust and intake ports should not exceed 56% of the cylinder diameter. Arch-shaped top and bottom edges of the channel ports help to keep the load on the piston rings due to the port edges, which increases as the speed increases, within bearable limits.

To reduce losses during gas exchange, the ports must be designed for effective flow. The design of the scavenging ports, in particular (**Figure 5.25**), has a great bearing on the engine output. They lead the unburnt gas charge from the crank chamber into the cylinder, deflecting it nearly 180°.

A port that is curved as evenly as possible, the so-called loop-shaped scavenging passage, shows the lowest flow losses. It has become common in compact engine designs, such as for chain saws. By using externally flooded pistons, it has been possible to move the lower orifice of the loop-shaped scavenging passage far enough upward that the cylinders can be short, stiff, and closed on the lower end. The externally flooded piston is made by retracting the piston wall in the area of the pin bores (**Figure 5.26**).

The process to be used in casting the cylinder is determined by the design of the scavenging ports.

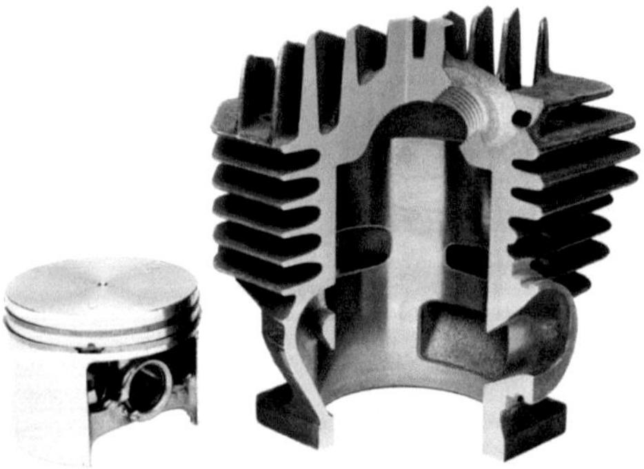

**Figure 5.26:**
NIKASIL® blind bore cylinder for a two-stroke chain saw: short loop-shaped scavenging passage with externally flooded piston

The most advantageous cylinders in terms of performance, with loop-shaped scavenging passages, require lost sand cores. This entails that they cannot be manufactured by high-pressure die casting. Composite cylinders with external scavenging ports that can be closed with a cover present an alternative solution.

Cylinders with straight, open scavenging ports–the reciprocating piston forms one wall of the port–can be manufactured by high-pressure die casting. The mixing conditions, however, are even less favorable than for closed, straight ports. Since this construction is not very stable in shape, it is suitable only for low-stress engines.

## 5.5.4 Cylinders for four-stroke engines

More stringent exhaust regulations are prompting engine manufacturers to pursue the four-stroke principle, even in small hand-held engines. Previous development efforts with the objective of making the designs as small as possible, lightweight, independent of orientation, and high-speed have not yet caught hold.

## 5.5.5 Surface treatment

Depending on the required service life and operational reliability, various running surface coatings are available to the user of light-alloy cylinders. In any case, light-alloy cylinders without running surface coatings or with thin reinforcements made of other metals are nearly equal in terms of heat dissipation and weight as well as less expensive compared with cast iron cylinder liners.

**Table 5.6** shows the ratings of light-alloy cylinders with various cylinder surface coatings in comparison with gray cast iron cylinders. According to the relationship

$$\tau = \frac{T^\circ \text{ gray cast iron}}{T^\circ \text{ light alloy}}$$

the temperatures at the base of the fin, near the combustion chamber floor, are compared for gray cast iron and light-alloy cylinders.

**Table 5.6:** Ratings of various running surface coatings, relative to gray cast iron cylinders

| Cylinder | Running surface coating | Rating of heat dissipation τ |
|---|---|---|
| Gray cast iron cylinder | – | 1 |
| Light-alloy cylinder with running surface coating | NIKASIL® | 1.16 – 1.19 |
| | CROMAL® | 1.16 – 1.19 |
| Light-alloy cylinders without reinforcement | – | 1.17 – 1.20 |

The following coatings have been proven in practice:

## NIKASIL®

The running surface of light-alloy cylinders coated with NIKASIL® consists of a nickel dispersion layer (**Figure 5.27**), about 0.05 mm thick in its finish-honed state, which has been developed by MAHLE.

This galvanically applied nickel layer, with an even inclusion of hard silicon carbide particles (<2.5 µm edge length), results in a relatively "smooth" cylinder surface after honing.

**Figure 5.28** shows the profile diagram of a typical NIKASIL® layer. The bearing roughness value TR has the magnitudes 0.4/96/1.2 (0.4 µm average roughness of the bearing surface with 96% bearing surface; average depth of deep scoring 1.2 µm).

NIKASIL®, as a cylinder surface, leads to fast run-in and associated rapid and good sealing of the piston rings. Together with the low friction power loss provided by NIKASIL®, optimal properties are obtained with regard to engine output, quantity of blow-by, oil consumption, and cylinder wear.

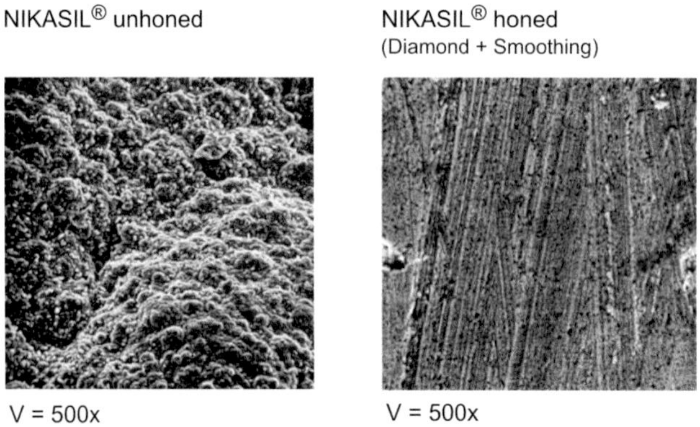


**Figure 5.27:** Running surfaces with NIKASIL®-coated cylinder, unhoned and honed

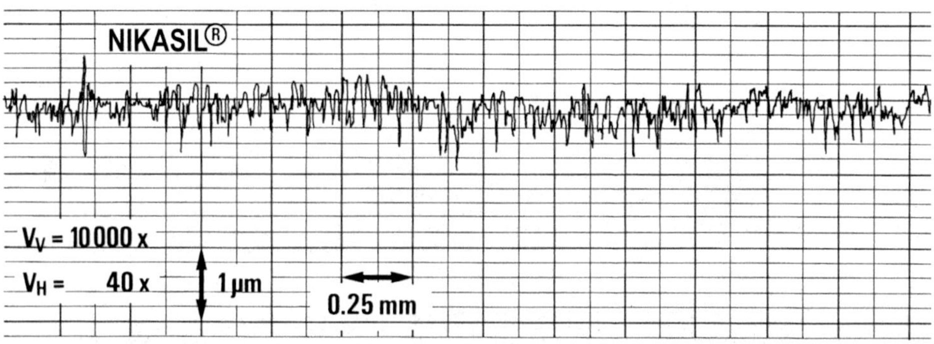


**Figure 5.28:** Roughness profile of a NIKASIL® layer

NIKASIL® cylinders are used in great quantities today, as standard, for air-cooled two-stroke and four-stroke engines. NIKASIL® is employed worldwide in racing engines, even for water-cooled four-stroke engines, with very great success.

## CROMAL®

This coating is a galvanically applied hard chrome layer of about 0.06 to 0.08 mm in thickness (**Figure 5.29**).

Wear of the chrome running surface and the piston rings that run against it are significantly affected in engine operation by the formation of the layer surface, which is intended to provide as even a distribution and adhesion of the lubricating oil to the cylinder surface as possible.

Honing with diamond strips, which is mainly used today, generates a roughness $R_a$ of about 2.5 μm, which is then smoothed to create a plateau structure.

The roughness profile in **Figure 5.30** shows a typical diamond-honed CROMAL® cylinder surface with a bearing roughness TR of 1.2/77/4.5. Honing with diamond strips has largely

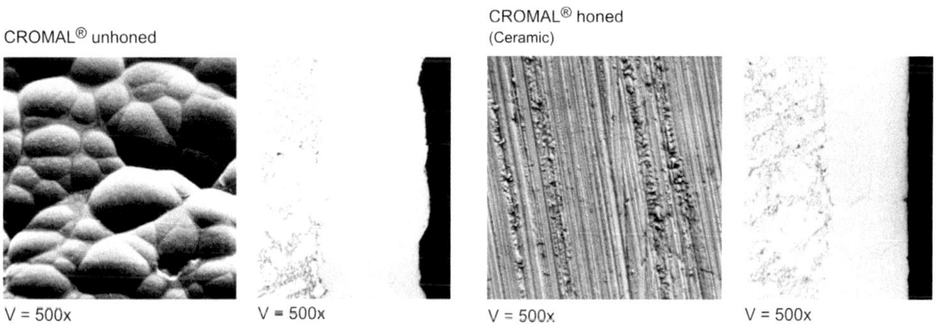


**Figure 5.29:** Running surfaces of CROMAL®-coated light-alloy cylinders, unhoned and honed

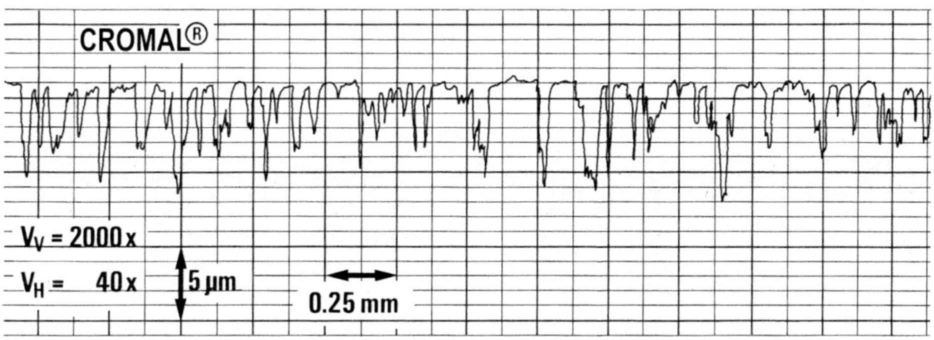


**Figure 5.30:** Roughness profile of a diamond-honed CROMAL® cylinder surface

replaced the previously used process of porous chrome plating or linked chrome plating as well as knurling. The correctly formed chrome running surface provides a good counterpart for the piston and the piston rings. The hardness and chemical resistance of the chrome coating lead to low wear values for the cylinder and piston rings and to a long service life.

# Glossary

| | |
|---|---|
| **1st piston ring** | First piston ring on the combustion chamber side |
| **Bearing clearance** | Gap between friction partners for the oil film to smooth out the loads |
| **Beveled ring** | 1st piston ring with beveled running surface on both sides |
| **Cast-in liner** | Cylinder liner cast into the engine block |
| **Cavitation** | Local material abrasion on the water-side cylinder wall due to implosions of water vapor bubbles |
| **Compression ring** | *See 1st piston ring* |
| **Conrod sweep** | Area covered by the connecting rod during a revolution of the crankshaft |
| **Cracking** | Splitting of the big end bore by fracture |
| **Cylinder liner** | Liner inserted in the engine block |
| **Cylinder surface** | Inner surface of the cylinder bore |
| **Double-beveled oil control ring** | Oil control ring with two running surfaces whose edges are equally chamfered |
| **Dry liner** | Cylinder liner that is cooled only indirectly by water, i.e., that does not have any direct contact with the coolant |
| **Expander ring** | Spring-loaded oil control ring |
| **Fixed-pin connecting rod** | Piston pin that is fixed in the small end bore by a shrink fit |
| **Flange bearing** | Radial bearing with axial thrust surfaces |
| **Gap clearance** | Spacing of the piston ring ends in the installed condition |
| **I-ring** | I-shaped oil control ring |
| **Keystone ring** | Piston ring with tapered side faces on one or both sides |
| **Lateral force** | Part of the force of combustion exerted on the piston that acts on the cylinder via the piston skirt |
| **Napier ring** | Special shape of a piston ring |
| **Normal force** | *see Lateral force* |
| **Oil control ring** | Piston ring designed to remove oil from the cylinder surface in a defined manner |
| **Piston pin** | Connecting member between the piston and connecting rod |
| **Piston ring** | Slotted, self-tensioning ring |

| | |
|---|---|
| **Rectangular ring** | Basic shape of $1^{st}$ piston ring |
| **Ring conformability** | Ability of a piston ring to conform to the cylinder surface |
| **Ring flutter** | Occurrence of radial or axial vibrations of the piston ring |
| **Ring gap** | End of the open piston ring |
| **Ring sticking** | Adhesion of the piston ring due to carbon buildup in the piston ring groove |
| **Running surface** | see *Cylinder surface* |
| **Scraper ring** | *see Oil control ring* |
| **Scuffing** | Piston ring face marks indicative of local overheating between the piston ring and the cylinder surface due to lack of oil |
| **Side faces** | Axial surfaces of the piston ring or piston ring grooves |
| **Thrust washer** | Axially acting bearing washer |
| **Twist** | Torsional deformation of a piston ring cross section due to a groove or chamfer on one ring side of the inner diameter |
| **Wet liner** | Cylinder liner washed with coolant from the outside |

# Keyword index

MIX
Papier aus verantwortungsvollen Quellen
Paper from responsible sources
FSC® C105338

If you have any concerns about our products,
you can contact us on
**ProductSafety@springernature.com**

In case Publisher is established outside the EU,
the EU authorized representative is:
**Springer Nature Customer Service Center GmbH**
**Europaplatz 3, 69115 Heidelberg, Germany**

Printed by Libri Plureos GmbH
in Hamburg, Germany